P9-CIS-391

BLOOD
AND
GUTS

BLOOD AND GUTS

A HISTORY OF SURGERY

Richard Hollingham

Foreword by Michael Mosley

Thomas Dunne Books
St. Martin's Press
New York

THOMAS DUNNE BOOKS.
An imprint of St. Martin's Press.

 For information, address St. Martin's Press,
175 Fifth Avenue, New York, N.Y. 10010.

www.thomasdunnebooks.com
www.stmartins.com

Library of Congress Cataloging-in-Publication Data

Hollingham, Richard.
Blood and guts : a history of surgery / Richard Hollingham ; foreword by Michael
Mosley. — 1st U.S. ed.
p. ; cm.
Includes bibliographical references and index.
ISBN 978-0-312-57546-5
1. Surgery—History. 2. Medical errors—History. I. Title.
[DNLM: 1. General Surgery—history. 2. Medical Errors—history. WO 11.1
H741b 2009]
RD19.H65 2009
617—dc22

2009031307

First published in Great Britain by BBC Books, an imprint of Ebury Publishing,
a Random House Group Company

First U.S. Edition: December 2009

10 9 8 7 6 5 4 3 2 1

To my mother, Penelope Ann Hollingham,
who would have made an excellent surgeon.

CONTENTS

FOREWORD

by Michael Mosley

In the early 1980s I trained to be a doctor at the Royal Free Hospital in Hampstead, London. I had five wonderful years, made lifelong friends and met my future wife, Clare. So although I now work in television and no longer do any form of hands-on medicine, I have few regrets about the years I spent poring over books and dissecting corpses. I did at one point in my training think about becoming a surgeon; after all it was a branch of medicine that was sexy, glamorous and well paid. Then something happened that made me realize that surgery was probably not for me.

One of the essential manual skills we had to learn early on was how to stitch up wounds. We practised by sewing bits of orange peel together and were then let loose on patients. The transition from uncomplaining oranges to human skin was always going to be a challenge. I remember with some embarrassment my first time. I was down in casualty on a Saturday night, a third-year medical student, intensely nervous. The place was crowded with the usual mix of drunks and minor injuries. I was asked to stitch up one of the drunks, an old tramp with a badly battered face, who had fallen over and gashed his forehead.

I pulled on gloves over my sweating hands and, with the assistance of a nurse, got together needle and thread and began to sew. I was slow, meticulous and careful. My patient was garrulous, confused and uncooperative. Finally I finished. But as I tried to pull away my left hand, which I had been using to hold the wound closed, we both got a nasty shock. I had sewn my glove to his head. I cut the stitches and started again, but I think I realized at that moment that I didn't have the manual dexterity, the precision, the sheer attention to detail that marks out the best surgeons.

Since then I have been in many operating theatres and watched many surgeons perform their magic. Fifteen years ago a surgeon saved the life of my son, Jack, and I have met many other people whose lives have been transformed by surgery. All this stimulated my interest both in surgery and its history, particularly the individuals and their discoveries who got us to where we are today.

The actual decision to make a television series rather than just think about it emerged from a conversation I had with Janice Hadlow, the dynamic controller of BBC4. I had just completed a series for BBC4 called *Medical Mavericks*, a history of medicine told through the stories of self-experimenters. Janice suggested that a series on surgery would be the next obvious thing to do. We soon agreed that the best approach would be a five-part series covering five different areas of surgery, and I went off to decide what exactly those programmes should contain.

Making television programmes is a collaborative process, and the end product the result of many different people's thoughts and insights. After some debate with my production team we decided to go for trauma surgery, cardiac surgery, plastic surgery, transplant surgery and neurosurgery. Each area illustrates something different about how surgery has progressed, and each is packed full of colourful characters and moral dilemmas.

We also decided that the programmes should not be purely historical but should start with an example of the best of modern surgery in that particular field. We would then use the modern case to look back at how the various elements of that particular operation had come about.

Many of the operations I witnessed while filming were memorable, but the one I found particularly striking was performed by cardiac surgeon Steven Westaby at the John Radcliffe Hospital in Oxford. The patient was thirty-four-year-old Sophie Clark.

Sophie had a couple of serious cardiovascular problems, which she'd had since birth. The first was a defect in a heart valve, the second an aortic aneurysm. An aneurysm is a weakening and swelling of a blood vessel, rather like a faulty tyre. As with a tyre, the risk is that under pressure it will burst. If the problem lies in the aorta, the main artery of the body, this would almost certainly mean death. The operation to correct both these two defects was extremely complex.

First Sophie was anaesthetized – a development pioneered in the mid-nineteenth century by William Morton, James Simpson and others (see Chapter 1).

Then she was connected to a heart-lung machine, the first of which was built and tested by John Gibbon in 1953 (see Chapter 2).

Next her heart was stopped, using potassium chloride, a chemical more commonly used for making fertilizer.

Then her body was cooled from a normal body temperature of 37°C to a decidedly chilly 16°C. This was to slow her metabolism and cut her brain's oxygen demands during the operation. It's an approach that was first suggested by Bill Bigelow, who was in turn inspired by research he had been doing into the hibernating habits of groundhogs (see Chapter 2).

Finally, all her blood was drained. As Steve put it, 'Heart surgeons are basically plumbers. You have to get the blood out the way just as you have to switch off the water before you change the pipes.'

At this point Sophie looked like something from the morgue. She was chilly to the touch, grey in the face, had no heartbeat, and the EEG technician could detect no signs of brain activity. She was as close to death as anyone I have ever seen.

Steve, under some pressure to get the operation done in as short a time as possible, did a magnificent job correcting her problems. He replaced her faulty heart valve with an artificial one, repositioned and reattached blood vessels using techniques first developed by Alexis Carrel (see Chapter 3), then warmed her up, started her heart, sewed her back up and the operation was done. She has since made a full recovery.

GOING FIRST

Not all surgery ends quite so happily. The thing about pioneering surgery is that it can, and often does, go wrong. The price of going first is that it is often those who come later who benefit from the lessons learnt. The history of surgery is littered with stories of patients who died while undergoing experimental procedures. In many cases, the sort of procedures attempted would not pass a modern ethical committee.

To be fair, the problem does not always lie with the surgical team. Take the case of Clint Hallam, the New Zealander who became the first man to have a 'successful' hand transplant. The operation took place in France in September 1998, and I remember vividly being impressed and slightly disturbed when I first saw this reported all over the news. I did not realize at the time that I would become involved and obsessed with Clint's story as it unfolded like a Greek tragedy.

In 1998 most people had got used to the idea of swapping body parts, as long as those parts were internal. Heart, liver, kidneys, lungs – all have long been eminently respectable organs to transplant, the main issues with them being around limited supply. Who should get the organ when it becomes available? Is it right to pay for organs? Questions like these were the main preoccupation. Suddenly we were confronted by something very different. Not only was the transplanted organ, the hand, quite obviously on display, but in some ways the operation itself was seen as 'cosmetic'. You can't live without a heart or lungs, and your quality of life without kidneys is poor. Surely, however, you can function perfectly well without a hand? The cost of keeping a transplanted organ is high. The drugs that prevent rejection will take something like ten years off your life. Many people felt that performing a hand transplant was morally indefensible.

Those who argued the counter-case – that a patient should have the right to choose whether thirty years with two biological hands was preferable to forty years with one – were not helped by what happened next. First it emerged that Clint had a criminal past (albeit for a minor tax fraud), then things began to go wrong with the transplanted hand. Clint stopped taking his pills and the arm started to be rejected.

At the time I was making a series for the BBC called *Superhuman*, looking at cutting-edge medicine. I sent a producer over to Perth, Western Australia, to film an interview with Clint. The transplanted hand looked absolutely terrible, more like a huge pink rubber glove than something human. It was useless for anything more sophisticated than holding a toothbrush, and it was clear that Clint now hated it. He talked about how people he met were repelled by it and said he was thinking about having it removed. However, he was still, in some wholly unrealistic way, also hoping to save it.

A year later I was flying back to London from California when I noticed Clint on the plane. We chatted about how things had been going and he told me that he was on his way to London to have the hand removed. It had reached the point where it was not just failing, but rotting. He had finally accepted that it was dead and the dream was gone. The following day he had it removed by surgeon Nadey Hakim.

So why did the world's first hand transplant go so badly wrong? When I asked Clint, he accepted that he had not been a model patient, but felt that his French medical and surgical team had not prepared him adequately for what was to come. In particular, he felt that the hand they had chosen to transplant was not well matched:

'I was ****ing angry with the doctors, and I am still angry that they didn't match it. It was huge and quite different to my other hand.'

Despite this, Clint told me that he does not regret having had the operation, and had recently rung around transplant surgeons offering himself as a candidate for a further hand operation. It's fair to say that there has been no rush to put him on a waiting list.

Since Clint's operation, more than thirty hands have been transplanted successfully. I went to Louisville in Kentucky to meet one of the most recent patients and try to understand what makes the difference between success and failure. On the way to the hospital I had a chat with my cab driver about his views on transplants. He felt that there should be no limits, that it should be down to the patient, the donor and the surgeon to decide. Oddly enough, he seemed most worried about where the donor organ had come from: 'I would not have an organ from anyone on death row as I would not like to have bad genes injected into my body.'

The surgeon who heads the transplant team at the Jewish Hospital in Louisville is Warren Breindenbach. Charming and hyperactive, Warren believes that what has been done so far is just the beginning; that eventually there will be no part of the body that's not transplantable.

Back in 1998 Warren and his team had been widely regarded as the ones most likely to perform the first hand transplant. Clint had travelled to Louisville and offered himself as a patient to the American team. That same year, on 23 September, Warren was in New York to meet Clint for further discussions when he turned on the television and discovered to his considerable surprise that not only was Clint in Paris, but he had had a hand transplant. The

French had got there first. When I asked Warren if he felt disappointed he said, 'I think every human being always wants to be a leader, but I have told my team and I have emphasized over and over again: it doesn't matter who does it first. It matters who does it best.'

Since 1998 Warren has performed three hand transplants, and the latest is perhaps the most remarkable. In November 2006 he led a team of surgeons in replacing the right hand of fifty-four-year-old David Savage. What is unusual about this particular case is that David had lost his hand in an industrial accident thirty-two years earlier. As Warren explained, this made the operation rather tricky: 'We ran into problems which were novel and new, and the analogy I make is kind of like closing your house down for thirty-two years, then coming back and deciding you are going to take a shower. You turn on the faucet and it sputters a little, and sometimes it works and sometimes it doesn't, so we had some sputtering as we tried to get the blood to flow into the hand that we were transplanting. But it worked.'

The operation was, in the end, a technical triumph. But I wondered if David, unlike Clint, was truly comfortable with his new hand. When I first met David and his wife, Karen, I was instantly struck by how different his new hand was from his other one. While David is powerfully built, relatively dark-skinned and has thick black hair on his forearms, the new hand was smaller, paler and more delicate.

I asked David if he found it strange to have the hand of someone now dead, and he said 'no'. Since the operation, he had felt it was part of him. I then asked him if he had considered the possibility that this hand had come from a woman, and he said he had, but it didn't bother him. He had found the fact that the fingernails on the new hand grow twice as fast as those on his own hand slightly disconcerting, but his main feelings were of gratitude to the family of the unknown donor.

David is undoubtedly happy to have had the operation, and optimistic about the future. When I watched him in his physiotherapy session I began to see why. He can catch things, lift up objects and manipulate tools. He has around 60 per cent of the function of a normal hand, and with more physio he may eventually get to 80 per cent.

The nerves that supply sensation are regrowing, and feeling is slowly coming back. He described with enormous satisfaction some of the simple pleasures of being able to use both hands again: 'Last September I went to my granddaughter's birthday party and just grabbing hold of her and picking her up was a fantastic feeling.'

Warren Breindenbach believes that successful surgery relies on cooperative patients. 'It's no good to only have a good surgeon. If you hook everything up properly but the patient goes home and doesn't use the hand, doesn't do physical therapy, then you get a lousy result. It is extremely important, the physical therapy and the cooperation, and that's where David has been an excellent patient.'

David's example made me question my initial scepticism about the benefits of this kind of surgery. I'm still not convinced that in his circumstances I would opt for a transplant, but I can certainly see why he did.

SELF-EXPERIMENTING

While making the series, I thought it would be interesting to immerse myself in the subject by repeating experiments from the early years of surgery. Things started innocuously enough with a spot of leeching. I had been puzzled as to why doctors and surgeons had gone on using bloodletting and leeches as a significant part of their practice well into the nineteenth century, so I met up with leech enthusiast Rory McCreadie, who promptly put a young, hungry leech on my arm.

After a few moments getting orientated, the leech bit and started to suck. Initially it was slightly painful, but then the bite went numb as the leech injected some form of local anaesthetic into the site. Rory and I now had to sit around for about an hour until the leech had gorged itself to the point where it was happy to let go. If you try to remove a leech before it is finished, it will leave its teeth behind inside you. Rory told me that in the eighteenth century leeches could be found all over England; a particularly good spot for them was Glastonbury in Somerset. Sadly, industrial pollution has wiped them out, and this particular one had been specially bred in sterile conditions on a leech farm.

Eventually the leech, now four times its original size, fell off. My blood then began to flow or, more accurately, slowly drip on to a plate. This is main point of the exercise. The leech injects an anti-coagulant to stop your blood drying up when it is feeding. After it stops feeding you continue to bleed; the aim is to lose about a cupful of blood. In my case, I went on bleeding for nearly twenty-four hours.

I actually found the whole experience surprisingly restful. Having experienced it, I can see that if the surgeon and his patient both believed in its benefits, then being leeched could have a powerful placebo effect. Unfortunately, it was a treatment with potentially serious side effects; some surgeons managed to bleed their patients to death. Examples include George Washington, who in December 1799 got a cold while out horse-riding. He died after his doctor repeatedly bled him, extracting in total around five pints of blood. I decided to stick to just the one leech, and took it home as a family pet.

My next experiment concerned pain. I wondered if any of the pain-relief treatments available to surgeons prior to the discovery of ether as an anaesthetic in 1846 would have been effective. So I went to a pub to find out.

First, I had myself hypnotized, then I tried sticking a needle through the web of my hand. It really hurt. Next I drank five double vodkas on an empty stomach and tried piercing my hand again. I felt supremely confident, right up to the moment when the needle went in. It was still really painful. Finally, I decided to try something a little bit more scientific: nitrous oxide. Also known as laughing gas, nitrous oxide had been widely used, mainly as a stimulant, since 1800.

A friendly anaesthetist arranged for me to try some nitrous oxide in the safe environment of an operating theatre. I took a few deep breaths and almost immediately began to feel the effects. I was intoxicated and euphoric. I was extremely pleased with myself and babbling with enthusiasm. But how good was nitrous oxide going to be at preventing pain?

My friendly anaesthetist had kindly brought along a device that violently stimulates the muscles of the forearm. When I had shaken off the effects of the first whiff of gas, I tried it. The result was painful and distinctly unpleasant. Next he gave me the highest dose of nitrous oxide he considered safe. Once again I went from being entirely sober to wildly intoxicated in a matter of seconds. When I was passed the pain-dealing device I seized it with enthusiasm. Once more my muscles twitched madly, but this time it wasn't painful at all; it was just funny. I could have gone on happily pressing the button for quite a while, but then the nitrous oxide wore off and suddenly my arm started to hurt and it wasn't funny any longer.

What I learnt from this bout of self-experimenting is that without the discovery of ether and then chloroform's anaesthetic qualities, it is unlikely that surgery could have progressed. None of the other options were powerful, consistent, long-lasting or safe enough to have allowed complex surgery to take place.

Few of the other things I tried out were quite as unpleasant as trying to force a needle through my hand, though immersion in near-freezing water came close. This was a re-enactment of an experiment first done by Bill Bigelow, whom we meet in Chapter 2, a pioneering researcher into hypothermia. Bill was convinced that if you cool down an animal, you can slow its metabolic rate and oxygen consumption. Do this with a cardiac patient, he reasoned, and you buy the surgeon more operating time. Rather than testing this idea on a dog, as Bill had done, the production team decided to test it on me.

So on a bitingly cold winter's morning I went for a dip in Hampstead Heath's men-only swimming pond. The temperature of the water was just above freezing as I arrived wearing trunks and some high-tech equipment. I had waterproof monitors to measure my heart rate and blood pressure. I also wore a mask that would measure my rate of oxygen consumption.

When I first went into the water it was extraordinarily painful, and I did quite a bit of whimpering. My pulse rate and blood pressure both doubled, while my consumption of oxygen also shot up. This was my body's instinctive response to the initial shock. After about five minutes my pulse rate and blood pressure had both fallen below my pre-immersion rates, but my oxygen consumption was still well above normal.

I had discovered what Bigelow also found in his dog: that cold induces violent, involuntary shivering, which increases oxygen demands. This was extremely bad news, as it meant that hypothermia would make operating on the heart more, not less, difficult. Bill persisted with his experiments, however, and soon found ways to abolish the shivering. When he did that, the animal's oxygen needs did indeed fall. Bill Bigelow's experiments led to the successful use of hypothermia in operating theatres, something I had witnessed at the John Radcliffe Hospital.

The application of cold, pain and leeches were all suitably historical, but I was also interested in trying rather more high-tech experiments. For the neurosurgery film, I thought it would be interesting to find out what it would feel like to have parts of my brain switched off.

For many years neuroscientists have known that different parts of the brain do different things, and that creating an accurate map of the brain is important for safe surgery (see Chapter 5).

In the early days doctors would find a patient who had had a brain injury, study what they could or could not do, wait till he or she died and then dissect their brain.

These days they have more sophisticated tools, which include transmagnetic stimulation (TMS). This involves using a powerful magnetic field to scramble brain cells temporarily in targeted parts of the brain. Having switched off that section of the brain, the scientists can deduce what it does by seeing what the volunteer is no longer able to do.

I wanted to see the effects of interfering with my motor cortex, the bit of the brain that governs fine movements, so I went to visit Dr Joe Devlin of University College London. It was a strange experience. When he turned on the TMS machine I completely lost control of the fine movement of my fingers. However hard I tried, I could no longer write, pour a glass of water or touch the tip of my nose with my finger. As soon as the machine was turned off, everything returned to normal.

This particular experiment made me reflect on how reliant we are on exquisite coordination between different parts of the brain and the body; how we only really appreciate what our body does when it no longer performs in the ways we expect. It is, of course, when things go wrong that we call first on the doctor, then on the surgeon.

We are extremely fortunate to live in an age when we have anaesthetics, antibiotics and machines for looking inside the body and the brain. We benefit enormously from the experiments and experiences of all those who went before. When I look back at what has been achieved in a comparatively short period of time by pioneers on both sides of the knife, I feel awe and immense gratitude.

In the course of researching and making the television series on which this book is based I met a lot of surgeons and their teams. I'm deeply impressed by what a varied, skilful, interesting and dedicated bunch of people they are, and I am very grateful for having had the opportunity to see them in action.

I'd especially like to thank Jonathan Hyde, Robert Marston, Ian Hutchinson, Alice Roberts and Peter Butler for their time and patience. Also Paulo Santoni Rugio, eighty years old and still doing facial reconstructive surgery in Cambodia.

I'd like to thank the production team at the BBC for their patience, insights and sheer hard work: Claudia Lewis, Kate Shiers and Kim Shillinglaw for driving the series editorially; Emma Jay, Giles Harrison, Hannah Liptrot, John Holdsworth and Sadie Holland for directing and producing the programmes; Giselle Corbett, Max Goldzweig, Sophie Guttner, Ruth Lacey, Fiona Marsh, Andrew Mayer, Laura Mulholland, Caroline Sellon and Sophie Wallace-Hadrill for providing the support work that made sure it all actually happened.

PREFACE

My mother always wanted me to be a surgeon. As a child, I spent more time hanging around hospitals than was probably normal. Mum was a nurse, so my sister and I became somewhat expert at navigating our way around the corridors of the Norfolk & Norwich Hospital. And, like most kids, I was often to be found in the Accident & Emergency department when I had gashed my leg, knocked my head in the playground, or swallowed a fish bone. I had my first operation when I was ten.

It was an operation on my right eye to correct a squint, and my family had absolute trust in the surgeon that it would be a success. Almost every moment, from being admitted to the hospital right up until the operation itself, stands out in my mind. I was in the children's ward in the old part of the hospital. It was reached through a long, chilly, stone-floored corridor. A few shabby partitions had done little to transform the ward from its Victorian origins. The ceilings were lofty, the radiators cast iron and the windows grimy.

I was given an injection before being taken to the operating theatre and remember examining the exciting cracks in the ceiling above my bed as the sedative took hold. I was wheeled along the corridor, up a ramp (which the porter had to make a run at) and to the operating theatre in the new part of the hospital. Outside the door of the theatre I was asked to count down from ten. I have no idea how far I got.

When I woke up my eye was covered with a bandage. The vision in the other eye was a blur. Someone brought me some ice cream. Three days later I was out of hospital and eventually went back to school. The operation was not a complete success. My right eye is still scarred (the scar is apparent when I get tired) and I had to have another series of operations a few years later, with a different surgeon, to correct the problem completely.

It is only when I look back at my first surgical experience that I wonder whether our trust in the surgeon was misplaced. Could he have done a better job? Was he having an off day? Was he desperate to get away for a round of golf? Surgery is risky, but we have somehow come to take it for granted that surgeons know what they are doing and that operations will be successful. However, even today, the decision to go 'under the knife' should not be taken casually. Imagine what it was like fifty years ago or even one hundred and fifty years ago.

In writing this book I've tried to re-create the surgical experiences of the past. It is a book about surgeons and the patients they operated on. Everything I have described really did take place and is based on accounts, reports, photographs, films and paintings from the relevant period. I have not had to exaggerate or sensationalize. In fact, in some cases I have had to tone down the stories to make them readable. I can assure you that the operating table at University College Hospital was stained with blood and that the operating theatre was next to the mortuary. Surgeons did inject paraffin wax under the skin, bombard patients with massive doses of radiation, and stick ice picks through their eye sockets – all in the interests of medical progress. There are some truly horrific episodes in the history of surgery that I have done my best to recount as accurately as possible.

There is, however, one important disclaimer. I came across the same problem encountered by the producers of the excellent BBC series that this book accompanies. There are so many stories that it proved impossible to include them all. As a result, this book is *a* history of surgery rather than *the* history of surgery. I have tried to include most of the more significant events, but also some of the most shocking, dramatic and entertaining. I have missed out whole areas of surgery, including orthopaedics and gynaecology, and some grisly early operations, such as those to remove bladder stones (you don't want to know). The chapters are arranged thematically rather than chronologically, which I hope makes the subject more accessible. I have also included a further reading section at the end to help you find out more. The only bits of the TV programmes that you won't find in the book are the presenter Michael Mosley's own contributions.

As you will have realized, I never did become a surgeon, journalism proving a much easier (albeit less lucrative) career path. However, I have for many years been fascinated by the history of medicine and surgery. One of my favourite TV programmes as a child was *Your Life in Their Hands*, when surgeons were shown performing real operations. One of my favourite museums is the Old Operating Theatre in London (see page 301).

Despite being immersed in the subject, there were certain events that even I found difficult to write about. Some of the accounts and pictures of injured soldiers and airmen, for instance, are deeply disturbing. I hope I have done justice to these remarkably brave men. I also hope I have given a fair account of some of the more controversial surgical treatments developed over the years, such as cross-circulation, lobotomy and brain implants. I am sure when you read this book you will find the stories equally compelling.

I owe an immense debt of gratitude to my wife, Susan, for putting up with me and for all her insightful editing and constructive criticism. I must also thank my mum, Penelope, who lent me a pile of books from her years in nursing and helped out with Chapter 1. As far as family goes, I also need to mention my son, Matthew, who was very patient with me at the Old Operating Theatre ('Is this it, Dad?' 'Yes, but isn't it fascinating?' Long pause. 'Can we go for a pizza now?'), and my father, Peter, who was the most recent Hollingham to go under the knife.

None of this would have been possible without the efforts of the programme production team, all of whom have been immensely helpful (their names are listed in the Foreword). They conducted much of the original research and, of course, have made some excellent TV programmes.

Thank you to the following people who helped me make sure I got my facts straight: Vivian Nutton from University College London (UCL); Alison Cook and Jonathan Hyde from the Royal College of Surgeons of England, and the various other surgeons Alison coerced into reading the drafts; Simon Chaplin, also from the Royal College of Surgeons, who put me right about Hunter; Peter Elliott from the Royal Air Force Museum, who helped me out on Spitfires and Wellingtons; Steven Wright from UCL, who provided a plan of Liston's hospital; and Stuart Carter, whose story is featured in Chapter 5. Finally, I would like to thank Martin Redfern and Christopher Tinker at BBC Books for their encouragement and support.

CHAPTER 1
BLOODY BEGINNINGS

OPERATING DAY

University College Hospital, London, May 1842

The operating theatre was positioned at the centre of the hospital, next to the mortuary. It was separated from the public areas by thick walls and a long corridor. This arrangement had two significant advantages: it helped shield passers-by from the screams; and its proximity to the mortuary meant that surgeons could move easily from operation to post-mortem, often with the same patient.

As it was, most people did their best to avoid the precincts of the hospital on operating days, and the staff did their utmost to distract anyone within screaming distance. It was not good for morale, particularly for those in the surgical wards who would soon go under the knife.

The steeply raked semicircular wooden galleries of the operating theatre had been swept that morning. The dust hung in the air, dancing in the few shafts of sunlight that managed to penetrate the grime of the high windows. A smoky coal fire burnt in a grate in the corner. At the centre of the room, where the surgeon would be performing, the gas lights hung from the ceiling on a chain above the operating table.

The table was made of deal – a cheap pine timber – and resembled a crude workbench. High and narrow, with a wedge-shaped block for the patient's head, it was bolted to the floor with thick iron brackets. The grain of the wood was marked with deep grooves and stained brown by the coagulated blood and soiled blankets of previous patients. Beneath the table was a box of sawdust, fresh that morning, although some still remained from previous operations, stuck to the side of the box like hardened brown putty.

One of the assistant surgeons, known in the hospital as a 'dresser', laid a thick woollen blanket on the operating table while his colleague carried in a case of surgical instruments. Both the men were nearing the end of their training and had already assisted in dozens of operations, although neither of them could say they had got used to it. The dresser carefully took the instruments from the deep velvet padding of the case. He laid them out in strict order on a tray placed on a small cabinet near by. He knew if he got the order wrong he would be in terrible trouble. He checked his notebook to make doubly sure.

Operating instruments:

- Two straight knives made of hardened steel, twelve inches long, with an embossed ebony handle and the sharpest of pointed blades

- A saw, short and polished, with fine sharp teeth and a good strong grip
- One pair of forceps
- Assorted sponges
- Threaded needles to tie blood vessels
- Short pliers or nippers to trim any jagged remnants of bone

The dresser covered the instruments with a cloth. There was also a bowl of water so that the surgeon could rinse the blood off his hands between operations.

Everything was ready. The first operation was scheduled to begin at noon.

In the male surgical ward the patient, rested and well fed, was as prepared as he would ever be. His bowels had been emptied that morning by means of an enema syringe, the resulting discharge being reported as 'copious and of bad quality at first' (the patient, the case notes recorded, was well rid of it). Two porters arrived to take the man to the operating theatre.

As they prepared to lift the patient from his bed on to a canvas stretcher, they could see that he was in a bad way. The poor man's lower leg had begun to suppurate: a thick fluid trickled from the open wound – a mixture of blood and pus seeping between the jagged ends of broken bone that protruded through the skin of his calf. The porters tried not to get too close. The smell of decay, like that of rotting meat, was almost more than they could bear. Without an operation the patient would die, that much was a certainty. The only cure for such a compound fracture was amputation, but with the infection creeping up the man's leg so fast that you could almost see it, the decision had been taken to remove his leg at the thigh.

The patient had sustained the injury on the Great Northern Railway when he had slipped between the platform and a moving train. Fortunately, the company's terminus at King's Cross was only a few hundred yards from University College Hospital. This meant he would be operated on by Britain's finest surgeon, Robert Liston. Liston had recently been appointed as the hospital's most senior surgeon, and professor of clinical surgery at the university. Author of the latest surgical textbook, he was the foremost surgeon of the age. And he knew it.

The room goes quiet as Liston strides through the door. 'Sharp features, sharp temper' is how his colleagues describe him. Most of his students (and many of his staff) are scared of Liston, but he is good at what he does and his operations are always well attended. True, there are those who have to attend them, such as the surgery undergraduates, but there are usually also rival surgeons and even visiting dignitaries in the audience. This is, after all, the very latest surgical practice, the best the British Empire has to offer.

Liston – six feet two inches tall, domineering and self-assured – hangs up his frock coat, takes an apron from the peg and rolls up his sleeves. 'Good afternoon, gentlemen,' he says to the now packed theatre. 'Today I shall be performing an amputation of the thigh in the usual manner.'

The two porters take this as their cue to carry in the patient. They lift him as gently as possible on to the operating table. The patient winces. This, he thinks, must be how condemned prisoners feel as they are led to the scaffold. His eyes dart around the room, his heart pounds, his utter terror mitigated only by the excruciating pain from his leg whenever he moves – a pain overlaid by a duller, steady, nauseating ache. He wants to vomit but can only gag.

Liston had made surgery his life's work and knew it had the power to save lives, but even he – described by his enemies as arrogant and aloof – operated only as a last resort. He also made every effort to instil in his students some sense of the feelings and fears of the patient. 'These operations must be set about with determination and completed rapidly, in order that dangerous effusion of blood may be prevented,' he told them. 'They are not to be undertaken without great consideration.' In short, it was all about speed.

The porters shut the doors and stand guard, arms folded, defying anyone to pass them. It has been known for patients to try to make a run for it, but this one only groans and mumbles indistinctly. It is probably a prayer. Most patients pray – it's amazing how many people find religion in the operating theatre. Many also beg or plead to be taken back to the ward even though they know that without surgery they will die. Others lie on the table calmly, as if possessed by some inner strength. Women, Liston finds, are often the most composed.

A dresser slips the strap of 'Petit's improved tourniquet' around the patient's upper thigh and pulls it through a small clamp. A wedge-shaped ridge on the strap is placed against the artery but not yet tightened, its purpose being to prevent blood loss during the operation. Without the tourniquet the patient's entire body would bleed dry in less than five minutes. Applying it properly was a matter of some skill. Tighten it too early and the upper leg would swell with blood. Too late and the patient could bleed to death.

Liston has himself witnessed the disastrous effects of a poorly applied tourniquet and is fond of telling the story in his lectures. 'A scene of indescribable and, under other circumstances, most laughable confusion ensued,' he says. 'Two assistant surgeons got on the table and pressed with all their might and main on the groin to stop the bleeding.' Liston is, in this instance, good enough not to reveal the surgeon responsible, but the story serves to remind people of his intolerance of error. The fate of the patient is not recorded.

With Liston in charge, there will be no such mistake today. One of the dressers takes a handkerchief from his pocket and ties the patient's good leg to the table to keep it as far away as possible from the knife. Two other assistants firmly hold the patient's shoulders and arms to stop him struggling. They try to keep their hands away from his mouth. He can squirm but cannot move; scream but not bite. The patient glances at the instruments, then at the ceiling. Finally at the audience – witnesses to his fate.

Liston motions for a young student to come forward from the gallery to support the limb that is going to be removed. The nervous pupil knows that if his grip slips or the leg bends, causing the bone to snap rather than be sawn through cleanly, he will suffer Liston's anger and abuse. He also hopes that Liston himself keeps a steady hand.

The surgeon clamps his left hand across the patient's thigh. His right hand reaches for his favourite knife, marked with a series of notches – one for each operation. The knife glitters in the flickering gaslight. Liston turns to the galleries; everyone is leaning over the railings to witness the action. 'Time me gentlemen!'

Those familiar with a Liston operation already have their pocket watches ready.

In one rapid movement, he slices into the flesh, and a dresser immediately screws down the tourniquet to stem the rhythmically spurting fountain of blood. Drawing the blade under the skin with the grain of the muscles, Liston pulls it towards the hip, down to the bone, then sweeps it around the leg and back towards the knee to leave two U-shaped incisions on the top and bottom of the thigh. There is nothing theatrical about the patient's cry. It is a chilling, horrible scream of terror. He is weeping now, struggling, mewling, whimpering.

Liston flings the knife into a tray and grabs the saw. His assistant puts his hand into the cut, fingers reaching right the way down to the bone. He pulls back the mass of skin, muscle, nerves and fat towards the hip to expose as much bone as possible. Liston places his left hand on the exposed bone and, with his right, begins to saw through it with rapid but precise strokes.

The student supporting the leg is concentrating so much that he barely realizes when he's holding its full weight. He looks down with a shudder, kicks the box of sawdust towards him and drops in the severed limb. It lands with a thud, sending up a small cloud of bloody sawdust.

The saw falls to the floor and, with his assistant still holding back the flesh of the stump, Liston bends close to tease out the main artery in the thigh – the femoral artery on the underside of the leg. The stump begins to ooze as Liston's bloodied hands reach for the needle and thread. He ties off the blood vessel with a reef knot. A 'good, honest, devilish tight and hard knot,' as he will later tell his students. He notices other, smaller, blood vessels and knots the ends together, holding the thread in his mouth at one point to make sure it is really tight.

Liston shouts at a dresser to loosen the tourniquet. A gently flowing stream of blood meanders between the ridges of the blanket to drip into a pool on the floor. But the pool is small, not large enough to be life threatening. The assistant allows the flesh he has pulled aside to spring back so that the bone is once again covered and protected by soft tissue. The two U-shaped flaps of skin are pulled together over the stump. A thin line of coagulating blood seeps between them.

The operation is over. From first cut to final stitch, the whole procedure has taken only thirty seconds. Thirty seconds of remarkable dexterity, flashing blades, rapid movements and brilliant showmanship. Thirty seconds of such pain that few patients are ever able to put it adequately into words. The memory of those thirty seconds will haunt them for the rest of their lives. If they live.

Fortunately, the mortality rate from Robert Liston's operations was remarkably good. Between 1835 and 1840 he conducted sixty-six amputations. Ten of his patients died – a death rate of around one in six. About a mile away at St Bartholomew's Hospital, surgeons were sending one in four patients to the mortuary, or 'dead house', where the all too frequent post-mortems took place.

Given that many surgeons were appointed through patronage or, more usually, nepotism, there was a large degree of surgical incompetence even in the most renowned hospitals. Surgeon William Lucas at Guy's Hospital in south London was generally kept away from the operating theatre for everyone's safety. In one thigh amputation he cut the U-shaped flaps of skin the wrong way round leaving a raw stump and a dismembered limb with two excess flaps of skin. His botched operations (the word 'botched' became synonymous with failed surgery) were notorious. They were thought to be the main reason that a young dresser at Guy's, John Keats, abandoned the surgical profession to become a poet.

In rural areas the local physician was expected to carry out his own operations. The medical literature of the day is littered with accounts of attempted surgical procedures and their consequences. Martin A. Evans, a physician in Galway, recorded a typical example from his casebook in the *Lancet* medical journal of 1834. His patient was forty-five-year-old Martin Conolly, whose leg was crushed by falling timber. Having persuaded the man that amputation offered the only chance of survival, Evans conducted the surgery, but his account gives little detail about the procedure itself, except that it was 'done by circular incision without assistance'. It is unlikely to have been as quick and efficient as Liston's operation, but was performed 'in the usual manner'.

As soon as the limb was removed Conolly reported feeling better and stronger, but in a few moments became faint and gradually weaker. 'He died,' reported Dr Evans, 'without having lost four ounces of blood during the entire process.' Evans attributes this not to any surgical failure resulting in massive internal bleeding, but to the patient. 'He had been a strong man, but was fearful of consequences, the only cause to which I can attribute his sudden dissolution.'

Patients had every reason to be fearful. Liston usually operated on reasonably fit young men or women with strong constitutions, and considered long operations cruel. That his were speedy affairs helped minimize blood loss and reduced the risk of disease. Liston also believed in keeping wounds clean. After the skin had been stitched together – with stitches known as 'sutures' (from the Latin word meaning 'to sew') – he advocated dressing the wound with sheets of lint dipped in cold water. These were to be frequently changed as the wound suppurated, with warm poultices applied to reduce the swelling and 'encourage discharge'.

Not for Liston the filthy bandages and straps of some of his rivals. These, as he was fond of saying, only encouraged 'putrefaction, fermentation, stench and filth'. It wasn't unusual for surgeons to reuse bandages and dressings already stiff with blood. For convenience, one surgeon proudly kept a drawer of 'plasters' passed from patient to patient over the years. Well, he and others reasoned, why waste them?

Liston would also wash his hands before operating and always wore a clean apron. Or at least it started off clean at the beginning of the day. Other surgeons took pride in conducting operations in the same frock coats they had used for years. The blood and pus that had built up into a hardened crust of material were regarded with respect. Surgeons were, after all, respected members of society; they had almost the same standing as doctors.

Liston and most of his contemporaries could, with some justification, claim to save lives. They had a firm grasp of anatomy, knowing with some certainty the name and position of every bone, muscle and organ in the body. They also knew broadly what each organ did, even if they had only a limited understanding of the underlying mechanisms. Crucially for Liston's generation of surgeons, they had also developed the skills and dexterity to stop their patients from bleeding to death on the operating table.

The decision to operate was determined by the pain the patient could withstand. In some quarters pain was seen as a prerequisite for a successful operation – a stimulant to the body's natural powers of recuperation. Perhaps the Galway patient had not been in enough pain? Many operations took far longer than the few seconds required for a basic amputation. Liston considered some of these too cruel. A mastectomy, for example, would take several minutes, the breast being slowly dissected 'with all due caution and deliberation'.

Neither was there any understanding of infection – what it was or how it was spread or prevented. Although Liston chose to operate in a clean apron with relatively clean hands, instruments and dressings, these practices owed everything to his sense of cleanliness and common sense rather than any theory of disease or how it was controlled.

The speed with which he conducted his operations, which included the removal of tumours and growths, and even reconstructive surgery (see page 217), was a hallmark of his work. Sometimes, though, his arrogance would get the better of him. (Indeed, the arrogance of surgeons is a theme throughout the history of surgery.)

Jealous rivals would whisper that Liston was so quick that he once accidentally amputated the penis of an amputee. On another occasion he was asked to look at a young boy with a swelling on his neck. A junior surgeon was convinced that the tumour was connected to the main artery in the neck – the carotid. 'Pooh!' said Liston as he drove a knife into the tumour. Unfortunately, the junior surgeon was right. The boy died within minutes.

However, the most worrying incident for his students occurred during an amputation when Liston accidentally amputated an assistant's fingers. The outcome of this operation was horrific: the patient died of infection, as did the assistant, and an observer died of shock. It was the only operation in surgical history with a 300 per cent mortality rate.

Liston's operations were messy, bloody and traumatic but, despite the occasional setback, he was one of the best surgeons of the day. His patients suffered terribly, but a fair proportion of them came out of hospital alive. This eminent surgeon owed his relative success to two thousand years of surgical development. A tortuous history involving dismembered criminals, wounded soldiers and Roman celebrities.

INSIDE THE BODY

Pergamum, Roman province of Asia Minor (Western Turkey), AD 157

The gladiatorial display was the zenith of Roman entertainment, a glamorous spectacle of skill, excitement and bloodshed. The day of the contest was one of celebration, and the amphitheatre was packed with expectant crowds ready to be entertained.

The day started with a display of exotic creatures gathered from the far reaches of the empire – leopards, wild horses and an angry bear. The animals were goaded in mock hunting demonstrations. A few were killed, but others were saved and employed as executioners to tear apart local criminals who were tied to stakes in front of the baying crowd. As the gladiators entered the ring, they waved to acknowledge the screams of the spectators, who idolized them as celebrities, their beautifully toned bodies admired by men and adored by women.

The gladiators fought in pairs – a warrior in heavy armour pitched against a nimble opponent with a net and trident; a fighter with swords against one with spears and daggers. Although the event was staged, the brutality of the fighting was terrifyingly – and thrillingly – real. The men fought to injure, to wound, to win. They were taught to aim for the arteries of the neck, and behind the knee. It was a fight to the death, but they shouldn't kill. The choice of whether a gladiator would live or die was the prerogative of the sponsor. He alone could decide whether the victor should deliver a final, fatal blow. The sponsor could not afford to allow too many gladiators to die – it would be like killing half the cast of actors after each performance of a play – as he would have to buy replacements.

Within the hierarchy of Roman society, gladiators were near the bottom of the heap. They were slaves and members of what was considered a disreputable profession. This was a standing they shared with prostitutes and, of course, actors. But despite their lack of freedom and their apparently low status, gladiators were rightly treated as the elite sportsmen they were. Their rigorous training was complemented by a high-energy diet and the very best medical treatment. The post of physician to the gladiators in Pergamum, or any major city of the empire, was a prestigious one. Celebrity gladiators required their own celebrity surgeon. This was the perfect position for a showman such as the ambitious Claudius Galen.*

Galen was a servant of the healing god Asclepius and had studied alongside distinguished physicians. This did not necessarily make him a surgeon, but he did know how to impress. When he was interviewed for the post of physician to the gladiators, Galen took along a monkey. He then proceeded to slice open its stomach and sew it back together again. 'Can anyone else do that?' he asked. He got the job and, as an added bonus, the monkey survived.

In his new role Galen would learn to deal with everything from minor sports injuries, such as muscle strains, to serious battle wounds. When the survivors of the contests left the arena Galen would be waiting to set bones or amputate limbs. He became an expert at stemming blood flow and restoring the fighters to health. As one of the first trauma surgeons, he was perfectly placed to study the inner workings of the human body. The exposed guts of a defeated gladiator, spilling out from a stomach wound, enabled him to examine the digestive system. An amputation revealed the bones, muscles and structure of the tendons, the bands of tissue that connect bone and muscle. He noticed how blood vessels pulsed and that some blood was brightly coloured. Galen later claimed that no gladiator under his medical care had died, but, even given their superior fitness, this is hard to believe. The physician, though, had a legend to build and a reputation to maintain.

* *No one seems to know for sure what Galen's first name was. 'Claudius' is used in many references, but some historians suggest it was more likely to be Aelius or Julius.*

As Galen's career advanced, he extended his studies of anatomy to animals – dead or, quite often, alive. He held public lectures and demonstrations where an animal was publicly dissected. Pigs seemed to bear the brunt of Galen's experiments as he considered them to be most similar to humans. His favourite demonstration involved severing the nerves in the neck of a live pig. As he cut them away, the wriggling animal became increasingly paralysed. First unable to move its hind legs, its front legs would then become still. With the final slice, Galen could stop the pig squealing.

In the manner of a true celebrity surgeon, Galen eventually became a personal physician to the emperor Marcus Aurelius. His ultimate ambition, though, was to become as famous as the Greek 'father of medicine' himself, Hippocrates. Galen hoped to be immortalized by the medical profession as someone who understood how the human body functioned. However, his only direct knowledge of human anatomy came from his work as a surgeon. Dissection of dead bodies was rare, and many considered it unclean and blasphemous. As a respected member of Roman society, Galen could not risk even suggesting such a thing, so instead he based most of his descriptions of human anatomy on what he had learnt by dissecting animals. The rest he surmised from consultations with his patients, or simply made up.

Much of what he deduced was right. Stopping a pig's squeals by severing its nerves made him realize that the brain controlled the voice. Aristotle had previously suggested that the brain was some sort of cooling system for the body. Galen concluded that arteries contained blood (rather than air) and that each organ had a particular function. He also advised that the strength, frequency and rhythm of the pulse could be used to diagnose disease. Indeed, he developed elaborate and complex theories on the differences between the various types of pulse that eventually stretched to sixteen books.

Some of his theories, however, were completely wrong. He taught that the blood was produced in the liver and distributed in veins. He saw the heart as some sort of furnace containing two chambers with tiny pores or micro-holes between them that allowed the blood to seep from one side to the other. There was no sense that the blood circulated around the body or was pumped from the heart. The pulsating movement of the arteries he attributed to their muscular structure, which he supposed contracted and expanded 'naturally'. And although he realized that urine was produced in the kidneys rather than the bladder, he got the position of the kidneys wrong.

Galen's crowning achievement was 'perfecting' the philosophical medical theories developed by the ancient Greeks: the four humours. Each humour corresponded to a different temperament and element: yellow bile was associated with fire; black bile with earth; phlegm with water; and blood with air. Illness occurred when the humours were out of balance. To rebalance the humours, the doctor could remove blood, induce vomiting or purge the body with an enema. A fever, for example, might be attributed to an excess of blood, so Galen advocated bloodletting to cool the body. A general feeling of melancholy suggested too much black bile, requiring the gut to be purged.

Galen believed he was a brilliant scientist and philosopher. Considering that most of his anatomical experience was based on animals, he did not do such a bad job. Many of his conclusions were based on real experimental evidence and would have made ideal foundations for later natural philosophers and doctors to refine and so improve our understanding of anatomy. The problem was that until the sixteenth century no one bothered.

The Roman Empire fell, Islam rose, Europe embarked on the Crusades, Columbus 'discovered' America, Magna Carta was signed and the printing press invented. Yet still, after one and a half thousand years, our knowledge of medicine, surgery and anatomy was based on the writings of Galen, a boastful Roman surgeon. That Galen was wrong about so much was hardly his fault, nevertheless it took more than one thousand years before doctors and surgeons began to question his teachings.

DEAD MEN'S SECRETS

Louvain, Flanders, 1536

It was nearing dusk. The city gates were about to be shut for the night. Outside the walls of Louvain, swinging on a gibbet in the gentle evening breeze, was the macabre silhouette of one of the city's criminals. The body was still more or less intact, but you could see through the ribcage. Ligaments connected many of the bones, but the skull was snapped unnaturally to one side – evidence that the hanged man had at least died quickly from a broken neck rather than from slow strangulation. Some parts of the body had fallen to the ground, the result of scavenging dogs jumping up and tearing them off. The kneecaps had gone, as had one of the feet.

Birds too had feasted on the decomposing flesh, their activities betrayed by the guano on the man's shoulder blades.

The authorities were particularly worried about undesirables arriving in Flanders from France. A decomposing corpse stationed outside the city gates sent a clear message that criminal activity would be severely punished. There was little evidence that the display worked, but it certainly unnerved most of the God-fearing citizens passing by.

This evening the road is deserted and a precocious medical student, Andreas Vesalius, is on his way home. He needs to be back in the city by the evening curfew, otherwise he will have to spend the night locked outside the city walls. He sees the gibbet at the roadside and goes across to take a closer look. The dangling corpse is exactly what he has been looking for – an ideal subject for study. Getting hold of bodies is difficult, and if he does not take this opportunity he thinks it likely that another medical student will.

The chain supporting the skeleton hangs some nine feet off the ground, and the corpse itself is considerably bigger than Vesalius. Even if he manages to get it down, he realizes he has little chance of carrying it home in one piece. With time against him, the obvious solution is to take one bit at a time, so Vesalius jumps up and grabs one of the legs. Tugging hard – it's surprising how strongly the femur is attached to the bones of the pelvis – he pulls it towards him until the ligaments rip with a hollow tearing sound and the joint pops out of its socket. He hauls it clear as the skeleton lurches to one side, dancing crazily on the end of its chain.

Vesalius then pulls off the second leg, twisting it out of its socket. The bones pile up at his feet. Next he decides to take the arms, but has to be careful not to damage the ligaments holding together the fragile bones of the hands. He also has to reach them. Fortunately, the wooden scaffold is rough and he can climb up, grabbing the swaying chain to pull the stinking corpse closer. Using one hand to steady himself, he grabs an arm with the other, gives it a sharp twist to pop it out of the joint and drops it to the ground. He then yanks the chain to spin the body round and reaches across for the second arm.

Jumping to the ground, he bundles up the bones in his cloak, like firewood, and hastens towards the city, keeping to the shadows as he makes his way home. Once there, he dumps the pile of bones on his kitchen table and, pausing only to pick up a hammer, heads back to the gibbet. He is determined to remove the rest of the body.

The head and trunk are all that remain, but the chain is attached to the top of the backbone and takes some hefty blows from the hammer before it comes free. Although he tries to catch them, these final body parts – the ribcage, pelvis and skull – fall to the ground, so he jumps down and wraps them in his cloak.

The gibbet casts a long shadow in the moonlight as Vesalius scrabbles around in the dirt to gather any stray pieces of cartilage and stuff them into his pockets.

The final challenge is to get back into the city. Night has fallen and the curfew is in place. Returning through the main gate would be foolish. Even an educated man such as Vesalius would have a hard job explaining his mission to the watch-keepers. He heads instead for another gate – one where he can slip past the guards unnoticed or, at worst, bribe them to get in.

Stealing a corpse from a gibbet might seem enough work for one night, but Vesalius is a man possessed. Safely back home, he now has a substantial collection of bones on his kitchen table – but they are starting to smell. There is more flesh on the bones than he thought. In the warmth of the kitchen, even the smoky fire can't disguise the sickly stench. Not only is this unpleasant, but the neighbours will notice. This could only lead to difficult questions. So, undeterred, Vesalius sets about stripping the corpse down to its bare bones.

After placing a large pan of water to heat on the fire, Vesalius gets a knife and scrapes away any last shreds of muscle, tendons and skin. He reaches his fingers into joints to separate out the cartilage and places this carefully to one side. When the water has boiled, he drops in the bones. He tries to keep the bones of some parts – such as the hands and feet – together as much as possible.

After a few minutes, he drains away the fat, straining the liquid to avoid losing any odd pieces of cartilage or fragments of bone. By daybreak the task is complete, the rotten flesh is discarded outside in the gutter and the bones and cartilage are gathered in an enormous pile on his kitchen table. Now all Vesalius has to do is put the skeleton back together again.

Although technically illegal, bodysnatching wasn't a wholly unusual pastime for a medical student. Few people were likely to complain if the remains of a criminal or pauper went missing. Frankly, it was doing everyone a favour. Neither was this the first time Vesalius had been involved in stealing a body. Desperate for some hands-on experience of anatomy, he had already joined with other young physicians to take bodies from the cemeteries of Paris when he was at medical school. But dead human bodies were difficult to come by, and Vesalius had a plan for his boiled-up bones.

In sixteenth-century Europe, Galenic medicine was as healthy as ever (unlike many of the patients who received treatment). Galen's 'scientific' writings had been recently rediscovered, translated and adapted; his theories re-examined and absorbed into the Christian doctrine for the modern European age. Medical treatment involved a thorough examination of the patient. The pulse would be read and therapy prescribed, depending on the imbalance of the humours. Just as in Galen's day if you went to the doctor with a fever there was a good chance he would get out the bleeding bowl or reach for the purgative and funnel to clear out your bowels.

But whereas the study of medicine was a respected, even noble, profession, surgeons were, as Vesalius would bemoan, considered little better than servants. There was, however, a growing fascination with anatomy. This was led, to a large extent, by artists. Renaissance artists were enthralled by the human body, its form, bone structure and musculature. And in the same way that doctors looked to Galen for insights, artists took inspiration from the beautiful statues of ancient Greek and Roman culture. A few years earlier, Leonardo da Vinci had become intrigued by the mechanics of the body. He produced intricately detailed drawings of human anatomy, of the brain, blood vessels and nervous system. Unfortunately, they remained unpublished during his lifetime.

Human dissection, almost always of criminals, was rare but gaining in popularity as part of medical training. Students were expected to attend lectures on human anatomy. At these, a professor would stand in a pulpit to read from Galen's text, while an assistant opened up the body. But these events could become awkward affairs when it became apparent that what Galen had described bore little relation to the anatomical reality of the human body. People were beginning to realize that, for all his genius, Galen had probably cut open animals rather than people. Some in the medical profession – particularly precocious medical students – were starting to question his wisdom. For surgery to develop, someone had to get a proper grip on where everything was and how it worked.

Vesalius set himself the task of reaching a fuller understanding of human anatomy. Back in his kitchen he began to sort out the bones and cartilage of the skeleton. He painstakingly identified each bone and laid it out in the correct position until his human jigsaw gradually came together. The parts that were missing, the foot and kneecaps, he 'obtained' from another corpse. There are 206 different bones in the human body, and Vesalius eventually laid out every one before carefully wiring them together into a skeleton that could be hung from a hook – not unlike the gibbet it was originally taken from.

The reconstructed skeleton was only the start. Over the next six years Vesalius dissected as many bodies as he could lay his hands on. Many were those of executed criminals; others he acquired from cemeteries. The contributions these dead people made to medicine were considerable. With their help, Vesalius was soon able to map every single organ, muscle and ligament in the human body.

Within the next few years Vesalius popularized dissection and started holding public anatomy demonstrations. These were attended by hundreds of spectators – not just medical students. Dissection became such a popular entertainment that the supply of bodies started to run out. This created a lucrative source of employment for less desirable elements of society. Working in gangs, bodysnatchers (or resurrectionists, as they would later be known in Victorian London) could make a comfortable income. However, the profession wasn't without its occupational hazards. Even if the authorities turned a blind eye, bodysnatching was still illegal. There was also the risk of picking up diseases. A small infected cut and you could soon be joining the other bodies destined for the dissecting table.

Vesalius published his work in *De Humani Corporis Fabrica* (The Construction of the Human Body). The invention of printing, using movable type and woodcuts, allowed him to include technically accurate and lavish illustrations. These pictures accurately identify the locations of all the major organs, nerves and muscles in the human body. The woodcuts show corpses posed in various unlikely situations, as if they are still alive. A picture revealing human muscles has the figure posed on a hillside in front of a town. There is a corpse dangling from a pulley and a skeleton resting against a tomb as if contemplating the meaning of life (or death).

The book was widely distributed and read by medical practitioners across Europe. In it Vesalius corrected more than two hundred of Galen's mistakes. These ranged from the structure of bones to the shape of the liver. In the second edition of his book, Vesalius also ruled out a connection (through Galen's micro-holes) between the two sides of the heart. However, even though he had worked out the structure of the heart, he still believed arteries originated in the heart and agreed with Galen that veins started in the liver. It was another eighty years before William Harvey concluded that the blood circulated around the body (see Chapter 2).

After 1300 years of stagnation, anatomy was finally on a firm scientific footing. Physicians and surgeons at last knew how the human body fitted together. Vesalius had broken the first barrier to the development of modern surgery. However, there were still three more barriers to go.

BLOOD ON THE BATTLEFIELD

A field near Turin, Italy, 1537

This is what happens when a musket shot hits a human body.

The bullet punctures the skin. As it does so, it drags fragments of clothing and gunpowder with it. The shot rips through the flesh, burning the tissue and splaying slivers of skin outwards. It gouges its way through the muscle, tearing apart the muscle fibres and severing tendons, veins and arteries.

As an artery wall is ruptured, blood starts to spray from the wound – pulsing at high pressure into the cavity the bullet has drilled. The bullet slows as it reaches the bone. The bone splinters, scattering sharp fragments. The two ends of the broken bone smash outwards through the skin. By now, the bullet has lost momentum and becomes lodged in the wound, mixing with the congealing bloody broth of muscle, bone, cloth and skin.

Injuries from musket bullets were far worse than anything that had been seen with daggers, swords or arrows. When a blade or arrow enters the body it inflicts a 'clean' wound and, with any luck, comes straight out again. But with the invention of the musket, and the larger guns that went with it, the battlefield was transformed. The few battlefield surgeons available had to cope with overwhelming casualties on a daily basis. When the guns opened fire and the men fell, the smoke mingled with a flume of fine red mist – blood spraying upwards from the injured and dying men.

Ambroise Paré had never seen such horror. The twenty-seven-year-old had been appointed as a battlefield surgeon to the French infantry commander at the siege of Turin. The army had been sent into northern Italy by the king, François I, in a long-running dispute over territory with the Holy Roman emperor, Charles V. By the time Paré arrived, the carnage was already horrific. To get close to the battlefield he had to ride across the bodies of dead and fatally wounded soldiers. Picking his way between them as best he could, he was forced to ignore their dying moans and pleas for help.

As Vesalius had noted, in sixteenth-century European medicine even experienced surgeons held little standing in society. The people who practised surgery were usually barbers. They had received no formal medical training and spent most of their time trimming beards or lopping off the odd wart. They might sometimes be employed to assist doctors with bloodletting. As for Paré, he was neither qualified nor registered as a surgeon. He had been working as a barber-surgeon at the largest hospital in Paris, he had no academic qualifications and no experience of anything other than the most basic surgical procedures. Everything he learnt about the profession (if it could even be described as such a thing) came from hands-on experience. He was familiar with basic anatomy and the theories of more advanced surgical techniques, such as amputation, but had not had the opportunity to put his knowledge into practice. He was going to have to learn fast.

Although the technology of war had advanced considerably over the last few centuries, battlefield surgery had changed very little. Surgeons had few options at their disposal. Any substantial wounds or compound fractures of a leg or arm usually meant the limb had to be removed. If a bullet entered a soldier's abdomen, surgeons might attempt to remove it with their fingers, or try to drain the wound of blood (and later pus as infection developed), but could do little else. For any seriously wounded soldier, the odds of survival were poor. However, surgery might give them a chance of life.

Every day Paré would saw off limbs. To stop the bleeding he used a hot cauterizing iron. As the leg or arm was removed, he placed the iron against the flesh – searing the muscle, blood vessels and skin together. Bullet wounds received the same treatment. With larger wounds, boiling oil was applied instead. Poured into the hole left by the bullet, the oil would burn everything it touched, destroying tissue but defeating blood flow. There was a belief at the time that gunpowder was poisonous, so cauterizing with an iron or pouring in hot oil had the secondary effect of destroying any poison. Or so the theory went. When bullets had failed to kill the soldiers, the shock of having boiling oil poured into their wounds often finished the job.

Cauterizing was not only brutal, it was also ineffective. By the time the surgeons had amputated a limb, a tremendous amount of blood had already been lost. Many soldiers bled to death before the arteries could be sealed shut. Even if they didn't die immediately, they would often lose so much blood that their chances of recovery became even slimmer.*

Paré started desperately looking for better and more humanitarian ways of treating battle wounds. His priority was to work out a more effective method of stemming the flow of blood. What little spare time he had was devoted to studying anatomy texts. When the guns went silent, he spent the evenings drawing diagrams and making reams of notes. His aim was to seal the arteries themselves – rather than the entire wound – block them off to prevent the worst of the blood loss.

His solutions were simple. His first invention he called a 'crow's beak'. The beak consisted of a set of curved forceps that could be clamped across the artery to block the flow of blood. Although other, smaller blood vessels would still be open, this device stopped the worst of the bleeding and bought him time during operations.

Next Paré devised a way of tying off blood vessels during amputations. This was not a completely new idea, but there is no evidence that it had been tried in practice before. Once the artery was clamped off using the crow's beak, he would tie off the vessel downstream of the forceps using silk thread. This 'ligature' would

* *Paré also had to deal with increasing numbers of casualties suffering severe burns. Lines of gunpowder would be laid by the enemy to create explosive walls of flame, cannons could misfire, and there were regular accidents with powder flasks and kegs. The salves available for burnt skin caused horrible blistering, and wounds often became infected as a result. Paré developed new treatments for burns and revised traditional ones. In one instance he used the juice of onions mixed with salt, which he applied to the wound with a cloth. He reported it as being a remarkable treatment.*

permanently block the artery. Starved of blood, the portion below the ligature would eventually die and drop off.

Paré published his first book, *Treatise on Gunshot Wounds,* in 1545. In it he detailed his experiences in combat and the lessons he had learnt. His practice of not using a cauterizing iron or boiling oil was widely adopted by those who read his work. The book revolutionized trauma surgery, or at least it did in many parts of mainland Europe. Unfortunately, because the book was written in French and not translated into Latin or English, other surgeons – particularly in Britain – continued to use cauterizing as a 'treatment'.

From a young, inexperienced, barely qualified surgeon, Paré went on to become one of France's most celebrated medical practitioners. His treatise was finally translated into English in 1617 as *The Method of Curing Wounds Made by Gun Shot (Also by arrows and darts).* The book is gloriously illustrated with a gruesome woodcut of a 'man of wounds'. The man has an axe through his head, a bullet through his leg and a dagger in his side, in addition to wounds from swords, arrows, spears and darts. Seventeen wounds in total. Even an accomplished surgeon like Paré would be hard pushed to treat him successfully.

Paré's crow's beak and ligature, although brilliant innovations, were less effective in practice. To stem the flow of blood completely following a thigh amputation, for example, more than fifty ligatures are required – although around ten would probably suffice to stop the worst of the bleeding. But in the dirt, smoke and poor light of a makeshift field hospital, even applying ten ligatures would prove completely impractical.

Likewise, trying to apply the crow's beak to a slippery artery that was spurting out blood at high pressure, while struggling to hold down a screaming patient, was an appalling challenge. It wasn't until the invention of an effective tourniquet (such as the 'Petit' type used by Liston) that ligatures really came into their own. But Ambroise Paré's contributions to modern surgery are nevertheless considerable. Above all, his efforts to reduce his patients' suffering shines through as a fine example to future generations of surgeons.

Thanks to Vesalius, Galen's mistakes had been corrected and surgeons now knew how the body fitted together. Paré had worked out how to tie off blood vessels and prevent patients from bleeding to death. What both men had in common was the courage to question the status quo; to challenge incorrect medical dogma. These were surgeons who trusted what they saw with their own eyes and learnt from their own experiences. Two major barriers to successful surgery had been broken. It would be more than three hundred years before the next major obstacle – pain – was overcome.

TWENTY-FIVE SECONDS

University College Hospital, London, 1846

Frederick Churchill of 37 Upper Harley Street was admitted to hospital on 23 November. Unmarried, and employed in service all his life, he had started as a footman and for the past sixteen years had worked as a butler.

A clerk noted down everything as the dresser asked a series of questions. The case notes would later run to some ten pages.

Aged thirty-six, Churchill was five feet eight inches tall with a fair complexion. His state of mind was cheerful and his sleep was generally sound. His habitual state of health was good, although not as strong as it had been eight or nine years ago. He was, the dresser noted, rather thin. Churchill's medical history included an attack of gonorrhoea eighteen years previously, and another attack around ten years after that.

In the year 1840 the patient had experienced a swelling in his right knee that became very painful. Severe pain was also experienced following a later fall in which the same knee was violently bent. In 1842 'considerably more' swelling and a 'discoloration of the leg ensued' following an injury to the left limb.

There had, the dresser recorded, been some outpatient treatment ordered by a medical man, but this had been discontinued. Then, in 1843, the swelling had been opened up – cut into with a knife – and 'a number of irregularly shaped bodies' were pressed out. These bodies appeared to have a fibrous, granular structure and varied in size from a pea to a large bean. They were preserved in alcohol and examined under a microscope. There were sufficient of these bodies to fill a two-ounce bottle.

'It is Professor Liston's opinion,' the dresser concluded, 'that these bodies are the remains of extravasated [forced out] blood.' Churchill's appearance was described as like that of someone in 'good but not robust health'. The right knee was much swollen and a probe could be passed through the cavity in the joint. Following this, Churchill was ordered to remain in bed. 'A thin serous discharge is given out. Pulse 80. Ordered to have a full diet and milk 1 pint.'

On 25 November Professor Liston examined the patient himself. He passed a probe into the knee and made an incision. Probing with his finger he could feel bare bone and the head of the tibia, one of the lower bones of the leg. He pulled on the bone to see if it was loose but this did not appear to be the case. Liston ordered that clean warm-water dressings should be applied and Churchill should undergo complete rest.

Churchill's condition began to deteriorate. He lost his appetite and the dressers noted that his tongue had become furred. More substantial food was ordered – a chop daily, a pint of beef tea and a pint of porter. On 27 November he experienced a terrible attack of pain extending from the hip to the toes. The swelling in the knee had increased and he suffered shivering, sickness and headache. Hot fomentations (poultices) were applied, which helped to relieve the pain.

On 17 December the dresser recorded that the patient 'had a kind of hysterical attack and was much excited', but by 20 December his appearance had improved and he appeared to be 'more healthy'. The next day he would go to the operating theatre to have the limb removed. Frederick Churchill had yet to be told that he would be part of a groundbreaking experiment.

At twenty-five minutes past two on the afternoon of 21 December, the porters carry Churchill into the operating theatre. As usual the galleries are filled with undergraduates nervously anticipating what was usually a dramatic, and often horrific, event.

Churchill is utterly terrified. He had known when he was admitted to hospital that it would probably come to this. At least with Professor Liston the ordeal would be over in a matter of seconds. Could he bear the pain? Could he appear strong in front of all these men?

Liston enters. The room goes quiet. 'We are going to try a Yankee dodge today, gentlemen, for making men insensible.' For the first time in the United Kingdom an amputation is about to be attempted using an anaesthetic. There had already been some trials at the hospital using hypnotism, or 'mesmerism', but the results had been mixed. Fundamentally, it was difficult to prove the scientific rationale for mesmerism, and among men of science it was considered superstitious nonsense.

The 'Yankee' Liston spoke of was the American inventor of the ether anaesthetic, William Morton, a Boston dentist. Morton had been trying a gas called ether – a pungent mixture of alcohol and sulphuric acid – on his patients during the extraction of teeth. (Given the generally poor state of dental health, there was no shortage of subjects.) Morton's process of 'insensibility' reached the attention of surgeons at the Massachusetts General Hospital in Boston, who were keen to use ether during operations. In a submission to the American Academy of Arts and Sciences, a surgeon at the hospital, Henry Bigelow, described the effects of ether both for dentistry and more serious operations. He reported a tooth extraction on a 'stout' boy of twelve. Upon wakening, the boy declared it was 'the best fun he ever saw'. The boy insisted on having another tooth extracted. In early November 1846 ether was tried on a young girl having her leg amputated above the knee. She lasted the whole operation without feeling a thing.

But Churchill doesn't know any of this. He lies on the operating table. A rubber tube is held to his mouth and he is told to breath through it for two to three minutes. The tube is connected to a flask containing ether gas. As Liston stands ready with his knife, the only sound in the room is Churchill's deep anxious breaths. Eventually the man becomes still.

After the rubber tube is removed from Churchill's mouth, a handkerchief laced with some drops of ether is laid over his face. Liston looks up at the galleries. The students are more excitable than usual – this would truly be one for the history books.

'Now gentlemen, time me!'

Liston slices his knife into Churchill's thigh. The tourniquet is tightened, Liston's swift movements cut the familiar U-shaped incisions, sweeping around the leg, pulling aside the flesh to expose the bone, to and fro with the saw, the ligature ready and the stitches in, the severed limb lying in a pool of congealing blood in the sawdust.

'How long, gentlemen?'

'Twenty-eight seconds.'

'Twenty-six seconds'

'No, I made it thirty!'

'Thirty?' exclaims Liston.

'Twenty-five seconds!'

This last figure has the time recorded in the case notes for the operation. Churchill has remained insensible throughout, not a sound came from his lips, not a groan, not even the slightest grimace.

'When are you going to begin?' exclaims the patient a few moments later.

This is greeted with peals of laughter from the gallery. There was rarely laughter after an operation. Churchill looks terrified. 'Take me back, I can't have it done!' Only when his amputated leg is held up for him to see does he believe that the operation has already taken place. He looks down to see his gently weeping stump. Later Churchill recalled feeling only a sense of great coldness and the memory of 'something like a wheel going round his leg'. The porters come forward with the stretcher to take him back to the ward. 'This Yankee dodge, gentlemen, beats mesmerism hollow!' declares Liston.

Later in the day another patient is given ether inhalation during an operation for an ingrowing toenail – previously an unbearably painful procedure. Flushed with success, Liston rushes off a quick letter to the *Lancet*, writing of the 'most perfect and satisfactory results'.

It is some minutes after Churchill is laid back in bed that he starts to feel any pain. By seven in the evening it has become excruciating. A dresser ties off more ligatures, making a total of ten altogether. Later, the two U-shaped flaps of skin are tied together with a series of sutures. Considering the agony, Churchill is remarkably cheerful, and as the evening progresses the pain begins to subside.

The patient is to remain in hospital for another seven weeks. On 31 December the dressers report that he is improving daily, the stump is healthy and 'discharging a small quantity of good pus'. A bandage is applied. By the end of January he is walking around on crutches. Frederick Churchill's case notes record that he was 'discharged, cured' on 11 February.

Soon, thanks to the pioneering efforts of a Boston dentist and Liston's reputation in Britain, almost every surgeon wanted to try ether. This Yankee dodge was surely the future of surgery. Some still felt that pain was an essential part of the healing process, but given the choice, what patient would want to go to *them* for an operation? During the Crimean War (1853–6), for instance, by which time anaesthetics were commonplace, surgeon John Hall was reported as saying, 'I like my patients to feel the smart of the knife.'

Liston took to holding parties at which ether was passed around the assembled guests. These social events usually included a cross-section of the capital's best-known artists and sportsmen, as well as surgeons, doctors and other gentlemen and their wives. Much hilarity ensued when the gas was tried out by some of Liston's assistants and they were seen to lapse into insensibility.

The relatively small doses of ether applied before operations meant that patients were 'under' for only a few minutes at most, yet the possibilities the successful relief of pain offered were endless. Operations no longer had to be so fast. Surgeons could take their time; they could attempt more complicated procedures. Robert Liston would not live to see the full potential of anaesthetics realized. He died in a sailing accident less than a year later. But by then his era of lightning-quick surgery was over.

THE MEDICAL STATE OF THE ART

While surgeons were saving lives with new techniques, medical science was struggling to catch up. The work of doctors had barely advanced since the Middle Ages, and if surgery was an inexact science, then Western medicine was more akin to a faith, a bit like astrology – scientific method built on foundations of sand. Treatments had changed little over the proceeding centuries and were limited in their scope. There were few cures available to doctors, and fewer genuinely effective drugs. Apothecaries boiled up all sorts of weird mixtures with varying results. A typical example from Guy's Hospital includes 'bath of herbs and sheep heads' prescribed to a woman suffering from an 'unknown illness'. How marinating the poor lady in offal was going to cure her was anyone's guess. Still, she probably paid handsomely for the privilege.

At best, all doctors could hope to do was to assist the natural process of healing. This might work for influenza, but would be completely ineffective against tuberculosis, syphilis or a heart condition. Even in the 1840s, the work of the physician was still firmly rooted in superstition. When you called on a doctor to attend you – and they did not come cheap – you might reasonably expect some sort of treatment. But the physician's options were limited. Medical practice was still based on the theory of the four humours developed by Galen. It was the job of the doctor to balance the bodily fluids of yellow bile, black bile, phlegm and blood.

As the understanding of anatomy had advanced over the centuries, most Victorian doctors knew this view of physiology no longer made sense. Yet the treatments available remained largely unchanged. Doctors could prescribe drugs. Some were effective for pain relief but others, such as mercury, were downright dangerous. Physicians would induce vomiting or diarrhoea in the patient to purge the body. They could also drain away excess blood. All these treatments made sense if you accepted the idea of the humours. They made no sense at all if you looked at the growing scientific evidence against them.

Bloodletting was as important to early Victorian medicine as it had been for almost two thousand years. Draining blood allowed doctors to remove 'morbid' matter from the bloodstream. This, the logic went, would be replaced by new healthy blood. Doctors carried scalpels or lancets (hence the name of the medical journal) to cut the skin and allow the blood to drain into shallow bowls. Others employed 'cupping' techniques, where small glass bowls were heated and placed over the lanced area of skin. The bowls cooled, forming a vacuum which helped to suck blood from the body.*

Some doctors preferred using leeches rather than cups. When leeches are attached to the skin they secrete a chemical that prevents the blood from clotting. This anticoagulant is so effective that even when the leech is removed, the wound will continue to bleed for another three to four hours. Leeches were particularly useful for bleeding sensitive areas of the body, such as the gums or around the eyes. American leeches were said to have a less irritating bite than British ones. The received wisdom was that leeches should be kept in a tub of river water with some peat or turf. It was best to rinse them before application.

For the modern physician who wanted to keep up with cutting-edge medical advances, scarifiers were the answer. These vicious contraptions resembled the mechanism of a clock and were marketed as the 'mechanical leech'. This 'new and modern' device contained a row of blades. When it was placed against the skin and a button was depressed, the blades sprang out to puncture the surface and induce bleeding.

* *Cupping also proved effective for pain relief, and was in common use in hospitals until the 1950s.*

The cases where bloodletting appeared to be effective were probably attributable to the placebo effect – the patient's belief in the treatment.* At least with bloodletting, patients were getting something for their money. However, by the 1860s evidence was mounting that the procedure was not only useless, but was probably doing more harm than good, particularly when advances in human physiology showed that bloodletting reduced the concentration of red blood cells. These contain haemoglobin, the protein complex that carries oxygen.

Nevertheless, and despite the scientific evidence, doctors were reluctant to abandon bloodletting altogether. At the turn of the twentieth century it remained a recommended treatment for high blood pressure (based on the 'common sense' argument that less blood meant less pressure). Even as late as the First World War, the technique of bloodletting was applied to the victims of gas attacks in the trenches.

Aside from blood, there were plenty of other bodily secretions to worry about. Urine in particular was seen as a valuable diagnostic tool. Not its chemical composition – its protein or sugar concentration – but its colour. Much, it was said, could be read into the colour of urine. Some specialists made their diagnosis on urine alone. Flasks of urine would be sent to them by other doctors for a specialist opinion. Often, somewhat inevitably, the prescribed treatment for the patient's ailment would be bloodletting. And so it went on.

* *There have been several studies over the years that show the placebo effect is actually quite an effective treatment. If patients believe a drug is doing them good, they tend to recover more quickly.*

Doctors seemed to be struggling to keep up with scientific developments. Surgeons, on the other hand, were as eager as ever to try something new.

MR SIMPSON CONDUCTS SOME INTERESTING EXPERIMENTS

Edinburgh, 1847

Ether was gaining in popularity, but the anaesthetic did have its drawbacks. It was a noxious gas to breathe, irritating the mouth and lungs. It had a tendency to induce vomiting in patients. The flask and tubes involved in administering it were awkward, and its effectiveness proved inconsistent. But the biggest problem was ether's high flammability.

Ether was being used only inches from the naked flames of the gas lights hanging over the operating table. The slightest upset and the gas was likely to explode in a ball of flame. There was also no way of telling how the prolonged use of ether would affect the patient. Would it leave them permanently unconscious or even brain damaged? Surgeons were used to their patients dying, but this seemed an especially unnatural way to go. There was also the question of its pedigree: it had been invented by a maverick 'Yankee' dentist. This was deeply unsettling to British, scientifically trained surgeons.

The only way the medical questions were going to be answered was to experiment on patients. Surgeons, of course, usually had no problem with this. Some, however, felt the drawbacks of ether were too great and started to look for an alternative.

James Simpson was a young professor of midwifery at Edinburgh University. As a student under Robert Liston, Simpson had attended his first operation aged just sixteen (he qualified in medicine at eighteen). The horror of the experience had lived with him ever since. Now head of obstetrics, he realized that every day he was witnessing more pain than ever.

Simpson was the son of a village baker, so to have risen to such a high position within the Scottish medical establishment was a remarkable achievement. He appears to have won the post through a combination of political persuasion (money may or may not have changed hands), public campaigning and, above all, an overwhelming sense of confidence and self-belief. It helped that he was also an excellent surgeon.

During a visit to London, shortly after the first ether operation, he had the opportunity to talk to Liston and confirm what was involved in the procedure. Perhaps he could apply pain relief during childbirth to relieve the terrible suffering some women had to endure? But questions about ether's safety become even more important when childbirth is involved. The gas would not only have to be used over a long period – hours possibly – but there was no knowing what effect it could have on the foetus. Might the child be killed or born an idiot? Simpson would also have to contend with religious and moral objections to the use of pain relief. Surely the pain of childbirth was a natural process? Didn't Genesis state that woman should bring forth her children in sorrow?

But Simpson was a driven man. He spent that summer trying out every chemical he could lay his hands on. He mixed a whole variety of substances together, drank and sniffed a cocktail of compounds. Every chemical that might prove a suitable candidate was inhaled or ingested. Then one day Simpson tried a new chemical that had been suggested by a Liverpool chemist. He woke up on the floor.

The last substance Simpson had tried before he passed out was known as chloroform. A colourless liquid composed of alcohol and chlorinated lime, chloroform had been invented some fifteen years earlier and marketed both as a treatment for asthma and a stimulant. That it had quite the opposite effect was an early cautionary lesson to not always believe what the pharmaceutical industry puts on the bottle. After trying the chemical a few more times, Simpson decided that chloroform needed rigorous testing before he used it on patients. So a few days later he took the opportunity to experiment on friends and family.

After dinner one night he served up tumblers of chloroform to some of the assembled guests. On breathing in its sweet, fruity aroma, they slipped into a magical state of relaxation. They laughed, they joked, the room started to spin, the conversation became distant and faint. The guests tumbled off their chairs or lay themselves down on the floor. Then everything went blank. 'This is better than ether!' exclaimed Simpson as he picked himself up off the rug some minutes later. 'A most pleasurable experience.' So pleasurable, in fact, that all the other guests were keen to try the thrill of chloroform for themselves. Simpson's niece had a sniff and presently declared herself an angel before passing out on the settee.

His scientific trial now complete, Simpson concluded that chloroform was a great success, and was in no doubt that it would bring untold benefits to his patients.

Four days later, Jane Carstairs is in the final stages of labour. Her screams following each contraction can be heard far beyond the delivery room. Bathed in a blanket of sweat, she is starting to become exhausted. Simpson knows he will have to intervene. He will probably have to use forceps, slipping the instrument – like a pair of long, wooden-handled serving spoons – either side of the infant's head. Then he will pull and it will hurt even more.

Simpson sprinkles a few drops of chloroform on to a handkerchief and lays it across Mrs Carstairs' mouth and nose. 'Keep breathing deeply,' he tells her. Within a minute she is asleep. When she awakes she is handed a little baby girl. The first success for this marvellous chemical.

This was the final proof Simpson needed that chloroform would transform nineteenth-century medicine. The charismatic surgeon saw it as his mission to spread the word. 'It was my duty,' he said, 'to teach all these people that they were wrong and I was right.' While taking every opportunity to try the drug out in his own practice (within a week he had used the new anaesthetic in an astonishing fifty cases), Simpson planned a marketing campaign to make sure as many people as possible knew about chloroform. Rather than publish his findings in a journal, couched in cautious scientific terms and possibly written in Latin, he took his results directly to the public. He drew up pamphlets that he sent out to other doctors. He gave talks and held demonstrations. He even took out an advert in the *Scotsman* newspaper, proclaiming this new miracle pain relief.

Not only was chloroform a more effective anaesthetic than ether, it was also a Scottish invention, and soon became a source of national pride. Simpson's confidence was infectious, and surgeons across Europe began to adopt his technique. Others sought to refine it, looking at new ways of administering the drug. One of Simpson's friends, a certain Dr Smith, tried to administer the drug rectally. Filling a syringe with chloroform, he injected it into his back passage. He woke up some hours later in a pool of diarrhoea with the syringe still in place, and suffered severe anal burns.

Apart from Dr Smith's misfortune, chloroform seemed to have few disadvantages. Patients were comfortable taking chloroform; it was easy to use and, unlike ether, involved no cumbersome equipment. As Simpson put it, 'No special kind of inhaler or instrument is necessary for its exhibition. A little of the liquid, diffused upon the interior of a hollow-shaped sponge, or a piece of linen or paper, and held over the mouth and nostrils, so as to be fully inhaled, generally suffices, in about a minute or two, to produce the desired effect.'

For the first few months everything seemed to be going well. Then, on 28 January 1848, fifteen-year-old Hannah Greener of Winlaton, near Newcastle upon Tyne, went to see surgeon Thomas Meggison for the removal of a toenail. She had undergone a similar procedure a few months earlier under the influence of ether, so was less fearful than she might otherwise have been. Nervous nevertheless, she was reassured by her uncle that everything would be fine. Mr Meggison would be using this new anaesthetic, chloroform. She would not feel a thing.

Hannah is seated in a chair. Meggison drips a teaspoon of chloroform on to a cloth and holds it to the girl's nose. She takes two deep breaths and pulls Meggison's hand away. He asks her to try again, this time breathing naturally. Half a minute later the muscles of Hannah's arm become rigid and her breath a little shorter. Meggison puts his hand on her pulse. It seems somewhat weaker but has not altered in frequency.

Meggison asks his assistant, Mr Lloyd, to begin the operation. Using a knife, Lloyd makes a semicircular incision and carefully prises off the toenail. Hannah starts to struggle and jerks forward. Meggison believes this is because the chloroform has not had sufficient effect, but he does not administer any more. Hannah's eyes are closed, but when the surgeon reaches forward to open them, they remain open. He starts to become concerned. Her mouth is also open, and her lips and face are suddenly pale.

Meggison calls for water and throws some in the girl's face. She does not move. He tries to give her some brandy. He holds it to her lips and, he later claims, she swallows – albeit with difficulty. Increasingly desperate, he lays her on the floor, cuts her arm with a lancet and tries to bleed her. When only a few drops come out, he tries bleeding her from the jugular vein in the neck, but manages to get only a spoonful of blood. Three minutes after Meggison administered the chloroform, Hannah Greener is dead.

An inquest before a jury was opened four days later and Meggison gave his account of Hannah's final minutes. The inquest heard from the doctors who conducted the post-mortem examination. They reported that the girl's lungs were in a 'very high state of congestion'. The coroner, J.M. Flavell, also included in the record an account of a chloroform experiment on mice. The mice had also died from congestion of the lungs. The jury concluded that Hannah 'died of congestion of the lung produced by chloroform'.

Simpson rejected the findings of the inquest, claiming that the death was more likely caused by the water and brandy. Subsequent studies have found it very unlikely that chloroform had a direct effect on Hannah's lungs. But it seems certain that it was at least partly to blame for the girl's death. And although she was the first to die under the influence of chloroform, she would not be the last. As surgeons started to use the drug for everything from ingrown toenails to major amputations, more and more people were dying. The deaths were sudden and dramatic as if, one surgeon reported, 'the patient had been shot'.

Curiously, the people who died were generally young and fit. Chloroform also seemed to kill a higher proportion of those who were more afraid of the procedure. Perhaps inevitably, deaths were higher in Scotland, where chloroform was the anaesthetic of choice, than in England, where ether was still preferred. Not that Simpson was experiencing any problems himself. All except one of his patients survived chloroform anaesthesia, but as the woman was quite frail anyway he was able to dismiss (in his own mind at least) any link to the chemical. A substance that had started out as a party trick, and was being used successfully on a daily basis, was also turning out to be a killer.

What the chloroform anaesthetic lacked up until now was any science. No proper studies had been done into what effect the chemical had on the body, or the doses that should be used. Simpson's advice to use 'a little of the liquid diffused upon a pocket handkerchief' was beginning to reveal its shortcomings. Should less be used for a young girl than an old man? How long should the handkerchief be held over the patient's face? These were fundamental scientific questions that no one had bothered to ask. Fortunately, someone else was already working on the problem.

In London Dr John Snow had been following the development of anaesthesia with great interest. He had already devised a number of improvements to administer ether, including a new type of vapour inhaler, and had drawn up tables to help surgeons calculate the correct (and safe) concentration of gas required.

Snow was the complete opposite of Simpson. He was a quiet, calm, diligent man given to careful scientific study. In 1848, as well as trying to save lives by improving anaesthetics, he was studying the outbreaks of cholera that were killing tens of thousands of people in the capital.*

In his publications on anaesthetics, Snow was very careful not to criticize Simpson 'in conferring on us the benefit of chloroform'. However, Snow was convinced that surgeons and doctors were using too much of it. He studied the effects of different concentrations of chloroform and divided them into 'degrees of narcotism'. In the first degree the patient was fully conscious, aware perhaps of the agreeable feelings felt when inhaling the chemical. The second to fourth degrees referred to various stages of insensibility or unconsciousness. Experiments with frogs suggested that a patient in the fifth degree of narcotism might stop breathing or suffer complete heart failure.

Snow concluded that chloroform had an effect on both respiration and the heart, and that there was a terribly fine line between

* *Snow worked out that cholera was spread through contaminated water rather than being carried in the air. Unfortunately, thousands more Londoners would die before his findings were accepted by the city authorities.*

insensibility and death. A third of a teaspoon of chloroform was enough to knock a patient out, but half a teaspoon could kill them. He reasoned that different people needed different doses. Young, fit patients might need more chloroform to render them unconscious, but this pushed them closer to a fatal dose. As for those who were 'fearful', it was probably because they were holding their breath for as long as possible. When they finally took a breath, they inhaled enough chloroform to stop their heart.

After applying his study to different classes and sensibilities, Snow took his conclusions further:

> Those persons whose mental faculties are most cultivated appear usually to retain their consciousness longest whilst inhaling chloroform and, on the other hand, certain navigators and other labourers, whom one occasionally meets with in the hospital, having the smallest possible amount of intelligence, often lose their consciousness, and get into a riotous drunken condition, almost as soon as they have begun to inhale. There is a widely different class of persons who also yield up their consciousness very readily, and get very soon into a dreaming condition when inhaling chloroform. I allude to hysterical females.
>
> *On Chloroform and Other Anaesthetics: Their Action and Administration* (1858)

Of course, the effect of chloroform had nothing to do with intelligence, educational attainment or class, but there was clearly some sense in controlling and regulating the dose. For those, such as 'hysterical females' who 'yield up their consciousness very readily', Snow advised using lower doses of chloroform. Whatever the subject's susceptibility to the drug, it was obvious to Snow that splashing a few drops on a handkerchief was downright dangerous. Just as he had devised better means of delivering ether to patients, he now set about designing an inhaler for chloroform.

A measured amount of liquid chloroform was added to a flask. This was attached to a tube, and a mask was placed over the patient's face. When the doctor cupped his hands around the flask this warmed the liquid, vaporizing some of the chloroform to a gas that the patient could comfortably breathe in. Even hysterical females. Snow's method was safe, easy to use and reliable. He administered chloroform to more than four thousand people. Only one of them died, and that was probably from other complications.

By 1853 Snow had become one of the physicians to Queen Victoria. During the birth of her eighth child, Prince Leopold, Snow administered chloroform. There were no complications with the birth and it is likely that he used only a small dose for pain relief. However, when the medical establishment found out, the *Lancet* published a furious editorial chastising Snow (although not by name) for putting Her Majesty's life at risk. The editorial spoke of the 'deplorable catastrophes' that were referable to the 'poisonous action' of chloroform, and the 'awful responsibility' of advising the administration of the drug to the queen.

Not that this controversy deterred Snow; after all, he had a genuinely fine track record. He employed chloroform again during the birth of Princess Beatrice four years later. Had Simpson been overseeing the birth, there might have been more cause for concern. That Snow, rather than Simpson, bore the brunt of criticism from the *Lancet* for risking lives with anaesthesia seems hardly fair. But then Snow was never to receive the recognition he deserved for any of his medical or public health achievements.

James Simpson died, aged fifty-nine, in 1870, a Scottish hero. He was the first man to be knighted for services to medicine. A huge state funeral was held in Edinburgh, the largest in Scottish history. Flags were flown at half-mast and thirty thousand mourners lined the streets. Statues and memorials were built. Hospitals were named after him.

John Snow, the man who had made Simpson's discovery safe, had died ten years previously. His great work on anaesthetics was published after his death. Snow's small grave was paid for by friends and colleagues. He also has a pub named after him.

Chloroform would continue to be used as a popular anaesthetic well into the twentieth century. In the end, Simpson's method of putting a few drops of chloroform on a piece of cloth became the most popular method of application. However, thanks to Snow's efforts surgeons used a lint mask and measured the chloroform using charts and a specially designed 'drop bottle'. Now (relatively) safe, chloroform could be used in the most difficult of circumstances and was the favoured anaesthetic in battlefield hospitals. Still, not everyone was convinced: some older surgeons were still suspicious of pain relief, preferring to hear the 'lusty screams' of soldiers as they went under the knife, a sure sign that the men were fighting for survival.

The invention of anaesthesia meant that surgeons had now conquered the third barrier to successful surgery. This, allied with a full understanding of anatomy and the ability to stem blood flow, meant they could now attempt new and more daring operations. People would seek treatment earlier. Women in particular might see a surgeon to have a small lump removed from their breast before the cancer took hold.

In theory, more lives than ever should have been saved. In practice, more and more people were dying. One out of five patients would probably end up in the dead house. In some hospitals, half of those operated on would be expected to die. Disease would ravage entire hospital wards. The disease even had a name: 'hospitalism'. Admittance to some hospitals amounted to a death sentence, and many people decided they would rather take their chances at home. Despite all the advances in science and medicine, no one could figure out why so many patients were dying.

NOW WASH YOUR HANDS

Vienna, 1846

Childbed or puerperal fever was a terrible disease. Within days of giving birth, the mother would start to experience discomfort, soreness and a rising temperature. Abscesses and sores developed and spread across the body, accompanied by a swelling of the abdomen. As the infection spread, it devoured tissues and attacked the vital organs. Meningitis – a swelling in the lining of the brain – might be accompanied by fits and periods of unconsciousness. Few women recovered. And while this was bad enough, in many cases their newborn babies died too.

In one clinic in the maternity wing at Vienna General Hospital, puerperal fever was killing hundreds of mothers each year. In January 1846, out of 336 births, there were 45 deaths. In February of the same year, 53 out of 293 women died. This was a death rate of 18 per cent – one in five patients.

There were two clinics at the hospital. In the First Clinic the patients were seen by doctors – mostly medical students. The Second Clinic was run entirely by midwives. When the authorities divided the maternity unit into the two clinics, they expected to see a rise in mortality rates in the ward where the midwives were in charge. It was, they argued, only common sense: midwives received less training, were less scientific in their approach and, of course, less rigorous in their intellect (there were no women doctors).

But the opposite was happening. In the clinic run by midwives there were far fewer deaths. In 1846 a total of 459 women (11.4 per cent) died in the doctor's clinic compared with 105 (2.7 per cent) in the midwives' care. It was a striking difference – and one that soon became well known throughout the city.

The two clinics admitted patients on alternate days. The changeover between the clinics was at four o'clock in the afternoon. Women in the advanced stages of labour would delay admission as long as possible so that they would be admitted to the midwives' rather than the doctors' clinic. As a result, women were giving birth in the street or in carriages. Others would run screaming from the hospital or had to be dragged through the corridors when they discovered they'd been put in the First Clinic.

Something needed to be done. A commission was set up to investigate the disparity. Its conclusions were desperate. Male student doctors, 'particularly foreigners', were blamed for being too rough in their examinations. Most foreign students were removed. When this failed to reduce the death rate the 'atmospheric-cosmic-terrestrial conditions of Vienna' were blamed for spreading disease – a 'miasma' was pervading the wards. The authorities struggled, though, to explain why, if there was something in the air, more women died in the First Clinic than the Second.

The patients themselves were blamed. These women were often the poorest in society; the wealthy would usually give birth at home (where the mortality rate was less than 1 per cent). Perhaps it was to do with the mothers' temperament or their slack morals? Many of them were fallen women. In an eventual admission of defeat, the authorities changed the days of admission, so women no longer knew which clinic they would end up in. It became, in effect, a lottery as to how likely they were to die.

There was one major difference between the clinics that the commission had failed to spot or perhaps considered unimportant. In order to refine their skills, the doctors had access to the bodies of the recently deceased. The midwives were forbidden by law to practise on cadavers, and had to make do with wax mannequins and porcelain models. As a result, the doctors and medical students spent much of their time in the mortuary. When needed, they returned to the wards to attend to their patients, the sweet smell of cadavers still on their hands. Some students even claimed that the scent was attractive to women.

In 1847 a twenty-nine-year-old Hungarian physician called Ignaz Semmelweis was appointed as first assistant to the professor of obstetrics at the hospital. He had responsibility for the First Clinic and saw for himself the horrible ravages of childbed fever. An intense yet kind young doctor, Semmelweis became obsessed with solving the mystery of all these deaths. Driven by the knowledge that for every ten patients he treated, two would die, he set out to find a solution.

As well as conducting autopsies on his deceased patients, Semmelweis pursued every theory he could think of. He suggested the disease was something to do with the position of the women when they were giving birth. Childbed fever seemed to affect first-time mothers more than others – perhaps this was something to do with their labour being more prolonged. Could it be fear of the doctors that was causing the deaths? Being examined for the instruction of male students was surely offending their modesty. If the women were already predisposed to puerperal fever, maybe their fear of being examined led to the onset of the disease? Of course, their modesty could be offended in many ways, so this theory was quickly dismissed.

Semmelweis observed that a priest was passing among the women – usually to administer the last rites. Maybe the disease was something to do with a man of the cloth spreading the fear of death? Certainly the priest had more cause to visit the First Clinic than the Second. He was very understanding when asked not to ring the little bell he carried around with him. But even if fear was a factor in the women's deaths, this would not explain the deaths of the infants.

Nothing seemed to work. Every theory Semmelweis came up with failed to answer the fundamental question: why were more women dying in the First Clinic than in the Second? Obsession turned to frustration and anger as he failed to solve the mystery. His superiors noted that he was behaving oddly, making lots of bizarre changes to little effect. Semmelweis needed a holiday – for everyone's sake.

In March 1847 he and two colleagues set off for Venice. The Italian city was part of the Austrian Empire and had, once again, become a popular tourist destination. This was in part because it was much easier to get to than it had been previously, thanks to the new railway line speeding through the Austrian countryside – a wonder of the age. It was hoped that seeing the art treasures of Venice would revive Semmelweis's spirits, and it did seem to have the desired effect. He headed back to Vienna reinvigorated, ready to resume the challenge of tackling childbed fever.

He returned to find that, in his absence, one of his best friends had died. Professor Jakob Kolletschka, a pioneer of forensic medicine, had become fascinated by finding out how people died, and conducted regular autopsies. It was during an autopsy that he met his fate. He had been dissecting a body with some students. The hand of one of them slipped while making an incision and accidentally pricked Kolletschka's finger. The professor thought nothing of it – the cut was small, these sorts of things happened all the time. For anyone involved in surgery or medicine, cutting yourself with a scalpel was an occupational hazard.

Within a few hours there was some redness around the wound, but nothing to worry about. The redness started to spread up Kolletschka's arm, he became feverish and sores began to develop. Soon he was covered in multiple abscesses and had a swollen abdomen. The post-mortem found that his organs were infected and he experienced pneumonia and meningitis. Kolletschka eventually became delirious and slipped into a coma. Only a few days after becoming infected, he was dead. Semmelweis was distraught.

Kolletschka was not only a close friend – the two men had often worked together, and Kolletschka had supported Semmelweis throughout his obsession with childbed fever. But Kolletschka was to help Semmelweis one last time.

Reading through the post-mortem protocol, it did not take long for Semmelweis to realize that his friend's symptoms were identical to those of the women who died of childbed fever. His mourning would have to wait. Now he knew what was killing the women. 'The exciting cause of Professor Kolletschka's death was known,' he proclaimed. 'It was the wound by the autopsy knife that had been contaminated by cadaverous particles. Not the wound, but the contamination of the wound by the cadaverous particles caused his death.'

Semmelweis had realized that if his friend had been killed by particles from a dead body, then the same particles were killing the women. Doctors were conducting autopsies and then administering to their patients. At best they might wash their hands with soap before conducting vaginal examinations, but this still left the lingering smell of the cadavers. The doctors were spreading the disease. They had been carrying death on their hands. Semmelweis had been killing the very patients he was trying to help. The conclusion was shocking. 'I have examined corpses to an extent equalled by few other obstetricians,' he wrote. 'Only God knows the number of women who descended prematurely into the grave because of me. None of us knew that we were causing the numerous deaths.'

Semmelweis decided that something more than a quick wash with soap and water was needed to stop the spread of material from cadavers to patients. In the middle of May 1847 he introduced a strict new regime in the clinic. Before examinations all doctors had to wash their hands in chloride of lime, a caustic chemical much like bleach. He posted notices to this effect:

> All students and doctors who enter the wards for the purpose of making an examination must wash and scrub their fingers and hands thoroughly in the solution of chlorinated lime placed in basins at the entrance to the wards. One disinfection is sufficient for one visit, but between the examination of each patient the hands must be washed with soap and water.

The results were better than Semmelweis could have hoped for.

In April 1847 there had been 57 deaths, the worst monthly mortality rate yet at 18.27 per cent. In May the figure came down to 36 deaths, or 12.24 per cent. The June figure was remarkable: there were only six deaths – a rate of 2.38 per cent, better even than the midwives' clinic. The following months were better still: in March and August 1848 not one patient died. Statistically, it was now safer for women to give birth in the hospital than at home. Thanks to Semmelweis, the hospital was now doing its job – saving lives.

The findings could not be clearer: childbed fever was caused by cadaverous particles transferred from the bodies of the dead. It was nothing to do with the atmospheric-cosmic-terrestrial conditions of Vienna, fear or foreigners. Semmelweis drew up tables to prove his point. 'Unchallengeable proof,' he said, 'for my opinion that childbed fever originates with the spread of animal-organic matter.'

He should have been a hero. Perhaps his manner did not help, or the fact that he himself was a foreigner. Some colleagues mocked him. They found the new regime of washing in chloride of lime inconvenient. It irritated their skin. And although everyone accepted that the number of deaths on the wards had dropped dramatically, where was the scientific explanation for Semmelweis's findings? What was this 'animal-organic matter' he talked of? How could the lingering smell of this material – this decaying flesh – possibly be enough to kill any healthy young woman?

His 'unchallengeable proof' was challenged by the head of the clinics – an ineffectual man drifting towards retirement. He did not want any controversy in his final months at the hospital, and this assistant was becoming increasingly troublesome. The doctors were complaining about this confounded new procedure. Semmelweis was sowing discontent and didn't know his place.

Semmelweis himself might also have been partly to blame for failing to get proper recognition for his work. He became entrenched in his views, would quarrel with anyone who disagreed with him and flew into rages. Except for those within the hospital and a small number of visiting doctors from elsewhere in Europe, few people knew of Semmelweis's discovery. Eventually his findings were published, but not by him. Some other hospitals adopted his procedures, but many did not.

The upshot was that hardly anyone outside the hospital and Semmelweis's immediate circle of friends knew anything of what he had achieved. His superiors eventually had enough of him and it came as little surprise when his contract was not renewed. In 1850 he returned to Hungary and took up the post of professor of obstetrics at St Rokus Hospital in Pest (later Budapest).

Here, if anything, conditions were even worse than they had been in Vienna. The Vienna General Hospital was at least a modern establishment, but of the eight beds in the obstetric unit at Rokus, one contained the dead body of a woman who had passed away the night before of childbed fever; the next bed contained a woman who was nearing the end of her life. The other six women were in the final stages of labour, but as they were also suffering from childbed fever it was extremely unlikely they would leave the hospital alive. The surgeon in charge carried out post-mortem examinations every morning before doing his rounds of the wards.

Semmelweis was quick to instigate a programme of cleanliness. He introduced chloride of lime and rigorous procedures for washing hands and instruments. By 1856 the mortality from childbed fever at the hospital was to drop to less than 1 per cent – lower than he had achieved in Vienna.

After much political wrangling (he was not the first choice for the job), Semmelweis was appointed head of obstetrics at the University of Pest medical school. The position sounded better than it was. The wards were filthy. The facilities consisted of a few cramped and poorly ventilated rooms in a tenement block. Of those women who were admitted – and even the poorest women made every effort to avoid this – a third would die of childbed fever. Again, Semmelweis introduced his reforms, but this time the mortality rate remained obstinately high. Then he examined the bedlinen.

In a cost-saving measure, the hospital had taken on a surprisingly cheap laundry firm. It soon became clear why the price was so surprising, when Semmelweis realized they were not actually washing the linen. They seemed merely to collect the stinking and stained sheets one day and return them in a similar condition the next. This time disease was being spread not by doctors but by 'matter' on the sheets. The laundry firm was sacked; the mortality rate dropped.

In 1857, now in his late thirties, Semmelweis married the nineteen-year-old daughter of a friend. His young wife gave birth to five children, the first died shortly after birth, the second from an infection – neither, at least, from childbed fever. Despite these tragedies, which were not an unusual occurrence in the nineteenth century, Semmelweis appeared settled and even declined a job offer from abroad. He decided it was time to write up his life's work.

When it was published in 1860 *The Etiology, Concept and Prophylaxis of Childbed Fever* was greeted with overwhelming apathy. Those responses it did receive were generally unfavourable; Semmelweis' theories were discounted. Prominent surgeons, including Scotland's James Simpson, lined up to criticize him. Many surgeons had theories of their own about the causes of childbed fever, including a suggestion that it was related to swelling of the Fallopian tubes. The Vienna General Hospital had already abandoned his 'crackpot' ideas as unworkable. 'We believe that this chlorine washing theory has long outlived its usefulness,' one doctor wrote in the *Viennese Medical Journal*. 'It is time we are no longer to be deceived by this theory.'

Unfortunately, even the most objective person reading the book would be inclined to treat Semmelweis with a degree of scepticism. Only a relatively small part of *The Etiology, Concept and Prophylaxis of Childbed Fever* is taken up with his experiments in Vienna and his defeat of disease. The rest reads as a bitter polemic on the way he was treated, underlined throughout by a sense of frustration that so few people would take him seriously. The epilogue reads as a morbid and futile cry for help, albeit tinged with some hope for the future.

> When, with my current convictions, I look into the past, I can endure the miseries to which I have been subjected only by looking at the same time into the future… If I am not allowed to see this fortunate time with my own eyes, therefore, my death will nevertheless be brightened by the conviction that sooner or later this time will inevitably arrive.

The rambling and sometimes vitriolic nature of the book revealed Semmelweis's declining mental health. He had become even more irritable, absent-minded and depressed. He wrote to doctors accusing them of murder for failing to listen to him. He went to the hospital chapel to pray for forgiveness for the deaths he had caused. He took to heavy drinking and visited prostitutes. His wife was being driven to despair. His own doctors suggested he take a holiday.

The Semmelweis family took a train to Vienna, where they were met by an old friend, Professor Hebra. The professor seemed keen to show Semmelweis his new hospital. Leaving his wife and children behind, he accompanied the professor to see the facilities. The hospital turned out to be the Lower-Austrian Mental Home. Semmelweis was held, tied into a straitjacket and confined to the ward for maniacs.

When his wife came to visit him the next day she was forbidden from seeing him. It seemed Semmelweis had tried to escape and had been restrained by six attendants. He was being held in a secure cell for his own protection. Accounts are confused about what happened next. Some believe that in being restrained he had in effect been beaten up; others say that he cut his finger (this could also have happened when he was restrained). Within days Semmelweis had become seriously ill: he was feverish, his body swollen, covered in abscesses and sores. Two weeks after entering the asylum, Ignaz Semmelweis was dead. He died from the same disease that had killed his friend Jakob Kolletschka and all those thousands of women.

At the time, few mourned his passing. They had always thought he was a madman, and the nature of his death only confirmed this belief. Anyway, as soon as he had entered the doors of the asylum Semmelweis was effectively dead, and his ideas with him. It was many years before his research was re-examined and his discovery – that cleanliness can prevent the spread of infection – fully appreciated.* But by that time, someone else had got the credit.

OPERATION SUCCESSFUL: PATIENT DIED

Glasgow, 1865

Joseph Lister was walking his ward. The professor of surgery stopped at each bed in turn to talk to the patients. He would ask the nurse to remove the sheets so that he could have a look at their wounds. It could be a depressing experience. The sweet, sickly smell

* *Decades after his death Semmelweis finally got the recognition he deserved. He now has a university named after him in Budapest, and is known by many as 'the saviour of mothers'.*

of putrefying flesh pervaded the room. Patients would arrive at the hospital in good spirits, confident of recovery. Two weeks later they would be dead. Despite all the advances in surgery, patients all too often succumbed to gangrene, fevers and blood poisoning. Any exposed wound was likely to become diseased. Even the most minor procedures, such as the removal of a small growth or wart, could end in a lingering death.

It saddened Lister how many amputations he had to carry out. A young child would come in with a fractured leg, having fallen awkwardly while playing or been knocked over by a cart or tram (there were increasing numbers of traffic accidents). If the child's skin was broken, Lister knew that within days infection would set in, the flesh would begin to rot and the limb would have to be amputated. Too many limbs were being lost because surgeons could not control disease.

Then there were the new operations that surgeons were trying to develop. They should be able to do more than hack off limbs or growths. With anaesthetics, they could take their time during surgery and try new techniques. Surgeons ought to be able to open up the body and operate on the organs. But, like any sensible surgeon, Lister would not operate unless absolutely necessary. He would not cut into the flesh unless he had to, and certainly not into the abdomen. Any wound was a potential source of disease and death.

The cause of this disease, or any disease, was still a mystery. Perhaps it was spread by bad air – some sort of miasma. But this was a new hospital. The wards had high ceilings, the beds were well spaced, there were large windows down the sides. With hard wooden floors and whitewashed walls, the wards were light and airy. Admittedly, the air was full of the smoke and smog of industrial Glasgow, but could that really be causing all this disease? Some of the more superstitious patients blamed it on the position of the hospital: it was built over the graves of cholera victims. Maybe they were right? Lister was prepared to consider anything.

Lister was a good but not particularly exceptional surgeon. He had risen gradually through the ranks to reach his position at Glasgow. Ever since he had been a medical student at University College London under Robert Liston, his overwhelming desire had been to save lives. Lister had witnessed Professor Liston's first use of anaesthetics and had closely followed advances in surgical techniques, as pain relief enabled surgeons to take more time with their operations. But the mortality rates from the amputations Lister carried out were still typical of the period. Around half the patients he operated on would die.

The development of surgery had ground to a halt. Surgeons knew how the body worked and they could control blood loss. They could even put their patients safely to sleep while they operated. Despite all this, far too many people who were admitted to hospital were dying. Until the problems of infection were solved, surgery could go no further. And opening up the abdomen to remove an appendix or operate on the organs was completely out of the question.

In his spare time Lister was also a scientist. There were few full-time scientists as such, apart from those in the chemical industry. For a gentleman, studying science wasn't really a vocation, more of a hobby. Science ran in the family. Lister's father, a wine merchant, was a respected microscopist and had devised significant refinements to microscopic technique. The younger Lister started his own experiments on frogs. He used a microscope to observe what happened when wounds became inflamed. He found that gangrene was a process of rotting – the flesh was decomposing. What he could not understand was why a simple fracture – a bone broken beneath the skin – healed, whereas a compound fracture – where the bone penetrated the skin and was exposed to the air – became infected.

One of the greatest tragedies in the history of medicine is how long it took the medical profession to realize that disease and infection were caused by micro-organisms. The invention of the microscope in the seventeenth century had revealed these 'germs' for the first time, but the work was never pursued and the connections never made between these 'microscopic' creatures and disease.

For all his achievements, not even Semmelweis had worked it out. He died believing that disease was spread by dead matter itself, rather than anything on the dead matter. Furthermore, few surgeons made the connection between dirty conditions and rates of infection. Florence Nightingale had shown how sanitary hospital conditions reduced death rates significantly, and even old-school surgeon Robert Liston had probably lost fewer patients than his rivals thanks to his attention to cleanliness. The fact that Liston operated so quickly also probably kept the death rate down. With anaesthetics, most operations were often taking longer, so wounds were exposed for a greater amount of time, increasing the opportunity for infection.

Doubtless more surgical patients survived in Victorian Britain than elsewhere, thanks to the obsession with order and cleanliness. But while most surgeons might be smartly turned out when they arrived at the hospital, when they came to operate they would don their old frock coat, encrusted with blood and pus – the result of years of messy surgery – and would pick up the same instruments they had used on the previous patient, wiped down to stop them rusting.

A professor of chemistry at the university, Thomas Anderson, told Lister about some experiments that had been conducted in France by Louis Pasteur. Lister found Pasteur's work simple but compelling. In one of his experiments, Pasteur sterilized a flask of broth by boiling it. He plugged the top of the glass vessel with cotton wool to allow the passage of air but nothing else. He left the flask for a few days and found the broth remained sterile. When the cotton wool was removed, the broth became putrid. Pasteur had proved that it was something in the air, not the air itself, that caused a substance to rot. The something, he surmised, was germs – micro-organisms in the air.

Pasteur's most famous refinement of this experiment was conducted using a swan-necked flask – a specially made glass container with a long, curved glass stem protruding from the top. Air could pass freely through the stem, but any dust or microscopic organisms in the air would become trapped. He filled the flask with broth and… it remained sterile.*

* *Pasteur's research was published in a series of papers between 1857 and 1860. Semmelweis was still working on his book during this period, but there is no evidence that he knew of Pasteur's work or that he ever made the connection between hospital infection and micro-organisms. Given that Semmelweis's achievements were published in only a very limited way, it is assumed by historians that Pasteur never came across his research.*

Reading through Pasteur's published research was heavy going, but Lister's efforts were rewarded. He started to piece together the evidence and began to realize what was happening to his patients: they were being killed by germs. So, he surmised, if he could kill the microscopic organisms or prevent them getting into wounds, there would be no infection. But Pasteur had sterilized his experiments using heat (a process that would later become known as pasteurization). How on earth could Lister sterilize a wound on a living person?

Lister tried a few experiments with various chemicals and compounds but with little success. The answer was to come from sewage. A hundred miles south, on the other side of the Scottish border, the authorities in Carlisle were trying out a new type of sewage treatment on the drains and cesspools of the city. The chemical they were using – carbolic acid – removed the terrible smell. Made from coal tar, carbolic acid had been shown in studies to kill germs. Lister reasoned that a chemical used to destroy micro-organisms in sewage might also be used to destroy micro-organisms in wounds and prevent infection. After all, the septic smell of rotting flesh pervading the surgical ward was not unlike that of untreated sewage. In the best traditions of surgery, Lister decided to try out his new 'antiseptic' principle on a patient.

On 12 August 1865, eleven-year-old James Greenlees was run over by a cart. He was admitted to the Glasgow Royal Infirmary later that day with a compound fracture of the left leg. The wheel of the cart had broken his tibia (the main bone of the lower leg) in two. The broken bone had punctured the skin, leaving a wound some one and a half inches long and three-quarters of an inch wide. When Lister examined the boy, he passed a metal probe into the wound to feel the broken bone. He observed that there was surprisingly little blood.

Under normal circumstances, the wound would have been covered and the boy rested. Splints would have been applied in the hope that the injury would heal, but Lister knew that eventually he would have little choice but to amputate. The boy would be left a cripple, his chances in life appallingly diminished.

Instead, Lister orders his house surgeon, Mr Macfee, to dress the wound using lint dipped in undiluted carbolic acid. The lint is laid across the wound and then covered with a sheet of tinfoil. The foil will prevent the carbolic acid from evaporating. Two wooden splints are then strapped on either side of James's broken leg.

Four days later, James says the wound is feeling sore, so Lister decides to take off the dressings to see what is happening. Urging the boy to keep still, he carefully removes the splints and peels back the dressings. Lister has never got used to this moment. Normally he would be forced to step backwards as his nostrils were hit by the smell of rotting flesh and putrefaction. This would normally be the time he would have to sit down and calmly tell the patient that amputation is the only option.

The final piece of lint is removed. He has never seen anything like it. There are no signs at all of suppuration; the wound is completely clean. The only smell is from the carbolic. The worst that could be said about the wound is that the edges are red – probably burnt, he thinks, by the acid. The soreness the boy has been complaining of is from the dressing, not, thank God, from disease.

Lister reapplies the lint, this time diluting the carbolic with clean water. Five days later he looks at the wound again – there is no pus or other sign of infection. However, the carbolic is still burning the skin, so Lister tries a mixture of carbolic and olive oil. After another few days he replaces this with a dressing of lint soaked in water. Six weeks later the wound is completely healed, the splints are removed and James walks home. It is, says Lister, 'a most encouraging result'.

By 16 March 1867, when the first results of Lister's work were published in the *Lancet*, he had treated a total of eleven patients using his new antiseptic method. Of those, only one had died, and that was through a complication that was nothing to do with Lister's wound-dressing technique.

Now, for the first time, patients with compound fractures were likely to leave the hospital with all their limbs intact. The next stage was to apply the technique to surgery. Operating theatres had changed little since Liston's day. The stained wooden operating table was usually surrounded by a raked gallery. When surgeons were operating the spectators would often gather close around the table, their outside boots grinding the dirt of the street into the timber floor. Light was provided by gas lamps or even candles. Devising an antiseptic operating technique under such conditions was quite a challenge, so Lister decided to rely on carbolic.

Before the operation he washes everything in a solution of carbolic. Hands, instruments, sponges and dressings are all dipped in the diluted acid. The patient's skin is brushed with carbolic, and towels soaked in carbolic are placed around the wound. To keep the air free of germs Lister employs a special contraption heated by a spirit lamp to send a spray of high-pressure carbolic steam over the operating table. The spray has to be adjusted to ensure the droplets are small because large ones could burn the eyes.

Once the patient has been put to sleep with chloroform, Lister rolls up his sleeves and the operation begins. The procedure takes place in a cloud of carbolic. Everything quickly becomes soaked. A fog covers the table and those surrounding it. Lister turns up the collar of his coat to avoid the acid reaching the skin of his neck. It is like operating in a rainstorm. When the time comes to close the wound, Lister uses sutures of catgut (made from the intestines of sheep) that have been soaked in carbolic. In the days of suppurating wounds it had been easy enough to pull out silk threads through the slush of decaying tissue. Now, as there is no infection, removing such sutures or ligatures could prove difficult. Not only is catgut sterile, but because the threads are organic, they are reabsorbed by the body and will not have to be removed later.

Operating under these conditions was deeply unpleasant, but the results spoke for themselves. Before antiseptic operations were introduced at the hospital, there were sixteen deaths in thirty-five surgical cases. Almost one in every two patients died. After antiseptic surgery was introduced in the summer of 1865, there were only six deaths in forty cases. The mortality rate had dropped from almost 50 per cent to around 15 per cent. It was a remarkable achievement.

Not everyone was so easily impressed. 'Listerism' was dismissed by some as nonsense. Despite the evidence, surgeons failed to accept the very idea of infection being caused by germs. They dismissed these 'little beasts' as a figment of Lister's imagination. Even those surgeons who understood the scientific basis for germs were not convinced by Lister's techniques. Operating under a spray of carbolic was inconvenient and unpleasant. New York surgeon William Halsted was even forced to operate in a tent because

Bellevue Hospital staff hated the fumes from carbolic so much. Other surgeons had been getting good results of their own simply by keeping their operating theatres clean and washing their hands properly. Lister rinsed his hands in carbolic but was still operating in his old, bloodstained coat.

Lister eventually abandoned the carbolic spray, realizing that there was a greater risk of infection from his hands or his instruments than from any germs in the air. It took more than ten years, but gradually Lister's ideas started to be adopted and operating theatres began to change. The rooms were scrubbed, the old wooden tables replaced by shiny metal, the floors sealed with linoleum. Surgeons hung their old operating coats up for the final time and started wearing clean linen shirts and operating gowns. They washed their hands and sterilized their instruments either by using heat or dousing them in carbolic. Wounds were covered with carbolic dressings. Some surgeons even started wearing rubber gloves. No one yet wore masks in the operating theatre, so a cough or a sneeze could still kill a patient, but death rates from operations continued to fall.

Listerism was here to stay and Joseph Lister became a national hero. He was the first surgeon to be awarded a peerage, and a public monument was erected in his honour. He even had a bacterium, *Listeria,* named after him, and thousands of people honour his memory every day when they gargle with Listerine mouthwash.

When Robert Liston, one of the world's finest surgeons, operated on patients in 1842 they had a one in six chance of coming out of hospital alive. If they had a compound fracture, an operation was their only chance of survival. For that they would have to endure the horrific torture of being held down on a hard wooden table, without anaesthetic, while their leg was sawn off. Ten years later they would have still have lost their leg, but at least there was pain relief and, assuming the chloroform did not kill them, a similar chance of survival.

Finally, by the end of the nineteenth century, surgery had become reasonably safe. The odds of survival had improved to better than one in ten (depending on the operation), and patients were much more likely to leave hospital with all their legs and arms intact. Despite many false starts, the four barriers to successful surgery had been overcome. Surgeons understood anatomy; they could stem blood loss and were able to control pain. Now they could even operate without causing infection. No part of the body was off limits. Surgery was becoming a science. Surgeons could do anything.

CHAPTER 2
AFFAIRS OF THE HEART

DOOR TO THE HEART

Montgomery, Alabama, 15 September 1902

It was only a few minutes after midnight and the small, dusty back roads of the city were pitch dark. The horse kicked up the dirt as it cantered along, the buggy jarring violently on unseen rocks and hidden potholes. Physicians of Dr Luther Leonidas Hill's reputation rarely came to this part of town; even during the day, it was easy to get lost. This was the negro area, where few could afford proper medical treatment; doctors usually called here only out of charity. But despite Hill's discomfort and the late hour, this house call was worth it. He might be able to save a life, if he was not too late.

Oil lamps burnt in the windows, and a small group of people had gathered around the door of the small wooden cabin. Several women were sobbing; another was trying to corral a group of bewildered-looking children. The older men hung back in the shadows, chewing on tobacco, mumbling and shaking their heads. A man and a woman stood by the door clasping each other's hands tightly.

Dr Parker and Dr Wilkerson were waiting outside and ushered Hill into the cabin's solitary room. The building was little more than a long shed, sparsely furnished with a table, hard wooden chairs and an iron stove in the corner. Then Hill saw the boy. Thirteen-year-old Henry Myrick was barely alive. He lay on the bed, his skin almost translucent, his breathing imperceptible.

Myrick had been stabbed with a knife at five o'clock that afternoon. The circumstances of the crime were not clear, but Hill found it hard to believe that the boy had been in a fight – his appearance was far too delicate for that. Dr Parker and Dr Wilkerson had been called six hours after the injury. Now, almost eight hours after the stabbing, Hill leant forward to examine the boy.

The knife blade had entered the boy's chest about a quarter of an inch to the right of the left nipple. Hill put his fingers to the wound and could see that it went deep. With every weak beat of the boy's heart, there was a bright red stream of blood, as if someone were squeezing a blood-soaked sponge. The skin was marked with a triangular patch of dullness, a bruise suggesting that most of the blood was being squeezed out inside the boy's chest. The boy's hands, lips and nose were cold. Hill felt for a pulse but could find hardly any sign of it. Even when he bent close, the heartbeat was barely audible.

The boy was slipping in and out of consciousness. Hill shook him gently and asked him how he felt. When the boy spoke, his voice was weak. He was clearly in great pain, but perhaps not beyond help. Hill went outside to consult Henry's parents. The surgeon was offering them a glimmer of hope that their son might survive. They agreed that Hill could operate.

When it came to the heart, Dr Luther Leonidas Hill, MD was one of the best-qualified doctors in the southern United States, if not the world. Obtaining his first medical degree at the age of nineteen, he had studied at medical schools in Alabama, New York and Philadelphia. From September 1883 until March 1884 Hill had even spent six months in London, being instructed by that great father of modern surgery Joseph Lister. Since returning to his native Montgomery, Hill had devised his own medical speciality: the study of heart wounds. Two years previously he had published a report drawing together the known cases of repairing a wounded heart. Hill had studied heart operations: he knew how they should be done and he knew how likely the patients were to survive. However, this was the first time he had seen a wounded heart for himself.

Hill asked for two lamps to be placed near the cabin's single table. The other doctors started to clean the area around it with carbolic as best they could. Hill lifted Henry from his bed and placed him on the hard surface. By now the cabin was becoming crowded with medical men. Hill's brother had arrived, as had a Dr Robinson, who was preparing to administer the chloroform anaesthetic. Hill doused his instruments in carbolic, then laid them out beside the table. Robinson took his dropper bottle, applied the measured amount of chloroform to the mask and held it over the boy's face. At one o'clock in the morning in a battered wooden cabin Dr Luther Hill was about to attempt one of the first operations on a beating human heart.

Hill raises his knife and makes his first cut through the skin to the left of the sternum, the breastbone that runs down the centre of the chest. It is a deep incision. He continues this cut outwards from the sternum along the third rib from the top. He must cut through the skin, the connective tissue beneath and the muscles covering the ribs. He makes a second incision along the sixth rib on the left-hand side, then joins these two lateral incisions together with a further vertical cut. Hill has carved three sides of a rectangle into the boy's flesh, the lines of incision now outlined in red as blood seeps out. This will become Hill's door to the heart.

Hill picks up some bone nippers and begins to cut through the three exposed ribs along the vertical incision in the boy's chest. The cutters go through the bone of each rib cleanly with a brittle snap. Ribs are attached to the sternum by pieces of cartilage, so Hill can lift up the skin where he has cut the bone and use the cartilage as a hinge. He gently pulls up the flap and bends it back, opening a door of skin and severed ribs to expose the heart.

The heart. The size of a large fist, this hollow muscular pump beats around seventy times every minute, 100,000 times a day, 36 million times a year. Over a normal lifetime the human heart will beat more than 2.5 billion times. Every minute it pumps some eight pints of blood around the body through more than 54,000 miles of blood vessels. Stop this circulation of blood for much more than four minutes and the lack of oxygen leads to permanent brain damage. Fail to repair a major wound or cut into the heart and a human can bleed to death within a minute.

Hill looks down at Henry's beating heart. Its protective fibrous sack – the pericardium – is bulging out, filling with blood from the wounded organ. The heart is struggling against the pressure of this blood pushing against it. The pericardium looks as if it could burst, and with every beat the situation is only getting worse. Hill slits the wall of the pericardium, enlarging the original stab wound. Blood pours out, but with the pressure on the heart released, the heartbeat grows stronger. This is a good sign.

Hill asks his brother to reach into the pericardium and pull the heart upwards towards the opening in the boy's chest. Finally, Hill can see where the wound has penetrated. The knife had cut through the thick wall of the left ventricle, one of the two long chambers of the heart. From the left ventricle oxygenated blood leaves the heart at high pressure to circulate through the body.

Hill's brother cups the beating heart in his hand. A jet of blood spurts from the wound with every pulse. It is difficult to keep the organ steady – the blood makes it slippery as it jumps in his palm, but he does his best. Dr Hill reaches for his curved suture needle and some catgut thread and begins to stitch the wound together. As he works, the flow of blood gradually lessens; the gap closes and the blood begins to coagulate.

The heart keeps beating.

His brother gently slips the heart back within the pericardium and Hill pours salt solution over it to both clean the cavity and act as a mild antiseptic. He closes the cartilage hinge and stitches the flap of bone, muscle and skin back in place. Forty-five minutes have elapsed and the operation is over. Hill lifts Henry back to the bed. The boy has a slight fever and is slipping in and out of consciousness, but his heartbeat remains strong.

Three days after the operation, Henry's condition starts to

improve. Fifteen days later he is allowed to sit up. Within a few weeks he has fully recovered and can show off his scar with pride. Hill is delighted; he is the first American surgeon to successfully cure a wound to the heart.

When Hill published a report of the case later that year, he included it in a table of similar operations undertaken between 1896 and 1902. For any aspiring heart surgeons the table would make depressing reading. There were wounds from knives, pistol shots and even scissors (in this case the victim had been stabbed a total of six times). Some of the patients received anaesthetic, some did not. Some were operated on immediately, some were not. It was difficult to draw any firm conclusions about the circumstances, given that so many of the operations resulted in death. Patients died of haemorrhages or infection, others bled to death on the operating table. Of the thirty-nine operations Hill had compiled, only fourteen patients survived (including the person stabbed with the scissors). In 1902 the chances were that two out of every three patients who underwent heart surgery would die. The odds were appalling. It was little wonder that most surgeons avoided operating on the heart altogether.

It is not as if the heart is particularly complicated. The organ is divided into two separate halves with a wall along the middle.* Vesalius (see Chapter 1) had accurately described the organ's anatomy in the sixteenth century, but believed blood was absorbed by the body and replaced by blood manufactured in the liver. In

* *Galen (see Chapter 1) believed this wall contained tiny holes that allowed the passage of blood from one side of the heart to the other. He was wrong, but before birth there is indeed a hole. This allows blood to bypass the lungs because they are not yet functioning. After birth this hole usually closes, although not in the case of babies born with a 'hole in the heart'.*

1628 (almost one hundred years after Vesalius) the English physician William Harvey published his essay entitled 'The Movement of the Heart and Blood in Animals', outlining his belief that the blood circulated around the body.

Harvey described how the heart is divided into two principal parts. The right side of the heart receives blood from the body and pumps it to the lungs; the left side of the heart receives blood from the lungs to pump it around the body. Each side has a smaller upper chamber called an atrium and a long, lower chamber called a ventricle.

Blood arrives at the heart through wide main veins known as the inferior and superior vena cava. This blood, low in oxygen, enters the heart and begins to fill the right atrium – a kind of holding chamber. When the right atrium is full, the muscle contracts to help push the blood through the tricuspid valve into the right ventricle – the pumping chamber. As the right ventricle contracts, blood is pumped out to the lungs to receive oxygen. This oxygenated blood returns to the heart in the left atrium, passes through the mitral (or bicuspid) valve and is pumped away from the heart in the left ventricle. The muscle wall of the left ventricle is thicker than that of the right as much more force is needed to push the blood all the way around the body. In a normal human heart, this whole process works smoothly and rhythmically: valves open and close, blood enters and leaves, muscles contract and relax.

The history of surgery suggests that surgeons have rarely been afraid of trying new, risky and untested procedures. By the 1900s surgeons were quite happy to cut into the body to operate on the internal organs. Appendectomies had become routine, tissue damage could be repaired and complex fractures set. Surgery was clean, relatively pain-free and generally successful. Surgeons were confident, highly respected members of society. But when it came to operating on the heart, they were terrified.

In 1896 the famous British surgeon Sir Stephen Paget declared that heart surgery had 'reached the limits set by nature'. More sobering to most surgeons perhaps were the words of Theodor Billroth, a pioneer of surgery on the digestive system. 'Any surgeon,' he wrote, 'who would attempt an operation on the heart should lose the respect of his colleagues.' And no surgeon wanted that.

Even twenty years later, during the First World War, surgeons would shy away from operating on the heart. Many soldiers had fragments of shrapnel left embedded in their chests, others simply bled to death. Some men survived for many years with bullets lodged in their hearts, the tissue healing around the foreign objects. Distinguished surgeon George Grey Turner summed up the situation when he was operating in a military hospital. He had the chance to remove a shell fragment, but concluded that 'it was beyond human and surgical capacity'. Even if the chances were that a patient would die without surgery, few surgeons were prepared to risk operating.

PURPLE HEARTS

D-Day, 6 June 1944

Within minutes of the first soldiers landing on the beaches of Normandy, the early casualties were on their way home. The landing craft became ambulances, shuttling backwards and forwards from the beaches to the ships. The ships went from being troop carriers to floating hospitals with makeshift wards and operating theatres. The walking wounded were patched up and returned to the beach to fight another day. As the ships wallowed in the heavy swell, the medics on board made every effort to keep the most severely injured alive. After the beaches had been taken and the Allied Army moved inland, the ships returned to England.

The military operation to evacuate injured soldiers was as well planned as the invasion itself. While hospital ships ploughed back and forth across the Channel, teams of nurses and doctors set up field hospitals to follow the advancing troops. There were flights to repatriate the wounded, fleets of ambulances, and even special hospital trains. Back in England, while the invasion force had been gathering along the south coast, land was being commandeered for new hospitals. The generals could only guess how many casualties were going to need treatment.

At Stowell Park, near Cirencester in Gloucestershire, the 160th United States General Army Hospital had only just been completed. Surrounded by the gently rolling Cotswold hills, lined with dry-stone walls and dotted with woodland, this was a perfect place to convalesce – it truly was England's green and pleasant land. Although the hospital was in countryside to avoid the aerial bombardment suffered by cities, it had good rail connections to London, the ports of Bristol and the south coast. An airfield had been built near by and an extensive network of concrete roads constructed across the site.

The hospital itself consisted of row upon row of Nissen huts – long sheds made of semicircular arcs of corrugated iron on brick bases. Some huts were wards, some were offices, some were operating theatres. There were mess halls and nurses' quarters and an officers' club. There was even a parade ground, not that the patients would be doing much parading. This hospital would receive some of the most serious casualties from the war – those men who would almost certainly have died in any previous conflict. The 160th General Army Hospital was the base for the Fifteenth Thoracic Centre and a daring, confident and ambitious young surgeon: the red-haired, Harvard-trained Major Dwight Harken.

Aged only thirty-four, Harken was held in high regard and was already shaping up to be one of the world's leading chest surgeons. By the time he came to lead the surgical team at Stowell Park, he had perfected new operations to remove cancers, and had worked alongside eminent surgeons in Boston (Massachusetts) and London before the war. He was convinced that no part of the body was off limits to surgeons – particularly the heart. What was the heart anyway but a mechanical pump? He could not understand why so many surgeons shied away from tackling heart injuries, instead allowing foreign bodies to remain lodged there – inhibiting the heart's function, dooming the soldiers to die a slow death from blood poisoning or, worse, triggering a sudden heart attack. Didn't surgeons have a duty to operate on the heart? Harken had lobbied his superiors, including the president of the Royal College of Surgeons, Grey Turner, to be allowed to carry out heart operations should the opportunity arise. Eventually, he was convincing enough to be given the go-ahead.*

As the first casualties began to arrive at the hospital, the nurses passed along the rows of stretchers, reassuring men that they would receive the best possible care. This was true, although some of the injuries were horrific. Some men were barely able to breathe, their lungs punctured by bullets. Others were coughing up blood or had chests swelling with fluid, their insides peppered with shrapnel.

Dwight Harken began operating around the clock, snatching sleep when he could, his energy and enthusiasm keeping him, and his team, going. Most of the operations involved opening up the chest to remove bullets, shrapnel and other debris – perhaps bits of uniform that had been in the way when the objects penetrated. It was remarkable that these men were still alive. Having entered battle at the peak of physical fitness probably saved them – that and the military effort to get them to hospital.

Harken also had technology on his side. Penicillin was now available, anaesthetics had been improved** and blood banks had been set up to enable transfusions (during the First World War doctors were still 'letting' blood for some injuries). Antiseptic

* *Grey Turner accepted Harken's reasons for wishing to operate, but added one more, telling the young surgeon that he had neglected an important consideration: 'namely, the knowledge of an individual that he harbours an unwelcome visitor in the citadel of his well-being'.*

** *Although in 1944 ether was still often used to induce anaesthesia, other options were now available to doctors, including injections of anaesthetic drugs. During major surgery, once the patient was 'under' a tube could be inserted directly into the trachea (windpipe) to pass air, oxygen or anaesthetic gases directly into the lungs. The mixture was controlled by the anaesthetist. Endotracheal intubation, as it is called, is still used today.*

technology had also moved on. The whitewashed operating theatre was kept as clean as possible. Everyone wore gowns and masks, and in addition to thoroughly scrubbing their hands, surgeons usually wore rubber gloves.

In addition to X-rays, Harken was also able to use a new type of imaging technology called fluoroscopy. This was much like taking a live X-ray image – X-rays were projected through the patient on to a fluorescent screen. Unlike the snapshot conventional X-rays provided, fluoroscopy could show a moving image. So day after day, night after night, images were taken, objects located and chests opened up. Lungs were stitched and reinflated, and infected tissue excised. When the men recovered, Harken presented them with the fragments of metal he had removed from their bodies.

But one day the fluoroscopic screen revealed a much more serious problem. The X-rays showed that the soldier had a bullet in his chest, but on the screen the bullet seemed to be jumping. There was only one conclusion – the bullet was lodged in the soldier's heart. With each beat, the bullet jumped. This was the chance Harken had been waiting for. He decided to operate.

Harken had prepared well. His previous experiments on animals had shown it was definitely possible to conduct delicate surgery on the heart. With each set of operations on dogs he was getting fewer and fewer deaths. He had a closely knit team of experienced doctors and nurses who were trained for this moment but, above all, he had the overwhelming belief that they were not going to fail.

The operating theatre took up half of a Nissen hut and it soon became very cramped. Around the operating table there were trolleys for instruments, the gas apparatus for the anaesthetic and a bulky electrocardiograph machine that drew an image of the patient's heartbeat on rolls of graph paper. Bottles of blood, matched to the patient, were brought in. As well as Harken, there were two other surgeons working as his assistants, an anaesthetist and a further surgeon to monitor the electrocardiograph. Alongside them was the scrub nurse, Shirley van Brackle. Everything was laid out ready; the young soldier was prepared for surgery and put to sleep.

Opening the chest is never something surgeons attempt lightly. There is so much that could go wrong. Soon though, Harken has made a foot-long incision and pulled apart the ribs with a retractor to expose the beating heart. It is obvious that there is a large fragment of foreign material in the right ventricle. Harken places sutures around the site, ready to be sewn together, then he cuts a small hole in the outer layer of muscle. Blood sprays out but the heart keeps beating. Can he stem the massive bleeding before the patient loses too much blood? And can he avoid the heart going into ventricular fibrillation – when the muscle loses its natural rhythm and beats uncontrollably?

Harken's hands are working in a well of blood; everything is bright red. He clamps his forceps firmly over the shrapnel and pulls. It sticks. The fragment of metal is plugging the hole he has cut. The bleeding stops; the heart keeps beating. Then suddenly, like the pop of a champagne cork, the object bursts out of the hole and so does the blood. It gushes in a torrent, a massive haemorrhage. The heart keeps beating, but time is running out. Harken has only seconds to close the hole before blood loss becomes too great. The patient's blood pressure drops but Harken doesn't panic.

As his assistant grasps the sutures in an attempt to tie them together, Harken flings the clamp and shrapnel across the room, narrowly missing Shirley the nurse. He makes another attempt to tie off the sutures, but nothing seems to stop the disastrous flow of blood. In desperation, Harken sticks his finger in the hole. The haemorrhaging stops, the heart keeps beating.

With his finger still in the hole, Harken begins to sew around it – underneath the finger and out the other side, gradually pulling the two sides of the gap together. One of the other surgeons jokes later that it would have been easier to cut Harken's finger off and leave it there, embedded in the heart wall. Harken slowly removes his finger as the sutures are tightened, but when he tries to pull his hand free, it will not come. He realizes he has sewn his glove to the heart wall. With the glove cut free, the blood pressure starts to rise. The soldier makes a complete recovery and Dwight Harken becomes the first surgeon to successfully cut into a beating heart. He is the first true cardiac surgeon.

Soon Harken would perform the operation again, and again, building up an impressive collection of trophies – shrapnel, bullets, fragments of clothing – all removed from soldiers' hearts. In the end he would operate on a total of 134 patients. There were no deaths. News of the remarkable surgery being undertaken at the 160th General Army Hospital soon spread. Everyone wanted to meet this dynamic young surgeon. There were visits from leading surgeons, generals, the Duchess of Kent, Queen Elizabeth (the future Queen Mother) and even Glenn Miller and his band, who played a few numbers in some of the wards. One of Harken's operations was made into a movie. As he worked, a Hollywood cameraman lay above the table on some makeshift scaffolding to capture the surgery in all its gory detail.

As the operations progressed, Harken gradually came to perfect his technique. New procedures were suggested and tried. At one point it was thought that an electromagnet might be useful to extract the metal fragments. It certainly seemed like a good idea, so a giant electromagnet duly arrived and was suspended over the operating table by a crane arrangement. Unfortunately, the implications of bringing a giant magnet into an operating theatre had not been fully thought through. When the switch was thrown and the magnet energized, the lights dimmed, the electrocardiograph went crazy and every metal surgical instrument in the operating theatre flew at high velocity towards it. Fortunately, there were no injuries but the idea was abandoned.

One of the surgeons who visited Harken was impressed by an operation, but questioned how much use this pioneering surgery would have after the war. What help would it be in peacetime to know how to remove bullets from a soldier's heart? But this visitor missed the point. Harken had done far more than perfect the removal of shrapnel. He had proved that it was possible to cut into a beating heart without killing the patient. The heart was no longer untouchable; it could be operated on safely and successfully.

Before the war intervened, Harken's ambition had been to operate on patients suffering from mitral stenosis. This disease affects the mitral valve, which controls the flow of blood between the left atrium and left ventricle. Mitral stenosis was usually the result of rheumatic fever and caused a narrowing in the opening of the valve. Sufferers from mitral stenosis endured all the usual problems of a weak heart, including poor circulation and breathlessness. The condition could leave them completely incapacitated and virtually guaranteed an early death. A couple of surgeons had attempted to cure the condition in the 1920s, leaving a succession of patients dead on the operating table. With his wartime experience behind him, Harken was ideally positioned to try again, and other surgeons had the same idea.

In 1948 Harken became one of four surgeons to successfully operate on heart valves.* Having proved that cutting into the heart was possible and survivable, the technique was relatively simple. The surgeons would make a small incision in the heart wall before inserting a tiny knife, scissors, or simply their finger to reopen the heart valve. They could not see the area they were operating on and had to feel what they were doing. All this would take place in a pool of blood while the heart was still beating.

It was known as 'closed-heart' surgery, although 'smash and grab' heart surgery would have been equally appropriate. As surgeons perfected their techniques, the procedure gradually became safer. Nevertheless, if there were any unforeseen complications, or a new experimental operation went wrong, the patient would usually die. And many patients did. Not only were the surgeons operating blind, they were also operating against the clock. With a hole cut into the heart, blood loss was tremendous. Although blood transfusions were used, surgeons had only around four minutes between cutting into the heart and sewing the hole

* *The first of these operations was carried out by Charles Bailey in Philadelphia. Harken carried out his first operation a few days later, but was the first to publish his results.*

closed before a fatal amount of blood was lost. Making anything other than a small hole in the heart would cause massive bleeding, and death would be virtually instantaneous. To attempt anything more ambitious, surgeons needed to see what they were doing and, above all, they needed more time.

DR BIGELOW AND THE GROUNDHOGS

Canadian prairie, near Toronto, 1951

Dr John McBirnie was having a miserable day. The prairie was bitterly cold, he was wet and up to his knees in dirt. Despite the fact that every farmer had told him there were groundhogs 'every-bloody-where' and they were a 'bloody menace', he had not seen a single one of the vicious bastards all day.

McBirnie didn't know what he was doing wrong. He had come well prepared for the role of chief groundhog catcher: he set off every morning dressed in waders and armed with a shovel, but his results were pathetic. He had tried digging them out and flushing them out with water. He had sat by their burrows; he had stamped up and down. Frankly, he was running out of ideas.

McBirnie had been assigned the job of catching groundhogs by Wilfred 'Bill' Bigelow, surgeon and director of the Cardiovascular Laboratory at the Banting Research Institute. Bigelow wanted to understand hibernation. In winter, when the prairie was covered in snow, groundhogs curled up in their burrows and hibernated. During hibernation the animals' core temperature cooled down to match their surroundings, their metabolism and circulation slowed, as did their heartbeat, allowing them to withstand temperatures only a few degrees above freezing. Bigelow had the idea of creating a similar state in humans – inducing hibernation to slow down the circulation. If he could reduce the amount of oxygen the body needed, perhaps this would buy surgeons enough time to be able to cut open the heart?

Bigelow had first got interested in studying the effects of cold in 1941, when he was a young surgeon at the Toronto General Hospital. His shift involved having to attend to a patient who had been drinking. The man had got so drunk that he passed out in the snow and when he woke up a few hours later, his hands were badly frostbitten. When he eventually got to the hospital there was not much Bigelow could do other than amputate the poor man's frozen (and now gangrenous) fingers. It was an unpleasant task, but the gruesome experience made the surgeon realize how little doctors knew about frostbite and the effects of cold. It inspired him to study how the body's metabolism reacts to low temperatures. Three years later he had published his first research paper on hypothermia.

After the war (and following a posting as a battle surgeon in the Canadian Army Medical Corps) Bigelow trained as a specialist in vascular and cardiac surgery. When he was working late one night he had a flash of inspiration. He realized that he might be able to apply what he had learnt about the cold to the problems of operating on the heart. He started to experiment on dogs.

The researchers immersed anaesthetized dogs in tanks of icy cold water to induce a state of hypothermia in an attempt to slow down the animals' circulation. The first results were baffling: the dogs were using up more oxygen when they were cooled than when they were at normal temperature. Bigelow realized that the dogs were shivering – even under anaesthetic. The muscle contractions were using up energy, so the muscles required more oxygen. But once the researchers switched to using ether anaesthetic – which also worked as a muscle relaxant – the dogs' temperature could be cooled by several degrees. With the animals' circulation and heartbeat slowed, the organs needed less oxygen. A 7-degree (Celsius) drop in temperature reduced oxygen consumption by half.

Bigelow was a generous man and openly shared his findings and published his results. Some thought he was mad; others thought the studies looked promising. The dog experiments had shown that the animals could be anaesthetized, cooled and their hearts operated on. When the dogs were revived, a good percentage survived and recovered well with no signs of permanent injury. This same technique might work with humans. Other surgeons started to take notice.

Meanwhile, Bigelow had a more ambitious goal in mind: he wanted to go beyond hypothermia to crack the secrets of hibernation. Could the research team find a chemical to slow down the body – a hormone perhaps? He set about collecting groundhogs. Or rather, because he was in charge, he made the (wise) decision to delegate.

Despite McBirnie's initial difficulties, Bigelow's team soon became adept at groundhog capture. They realized that the best way to get the animals out of their burrows was to flush them out with water. Three trucks moved from farm to farm, a line of spectators in their wake, as farmers and other locals came to watch. It seemed that this was the most exciting thing that had happened around these parts for a long time.

The first truck was the scout car; the scout car team was responsible for finding the groundhog burrows. The next vehicle was a tank truck full of water. Bringing up the rear was the truck carrying cages. Once the animals were captured they did everything they could to escape – they chewed through chew-proof cages, they escaped from escape-proof containers, the sharp-toothed little brutes would bite researchers' hands as a matter of course. Some members of the team began to dread the work. All of them came to treat the animals with great respect.

Eventually Bigelow had enough groundhogs to establish the world's first (and only) groundhog farm. A large, fenced-off field, complete with luxury (in groundhog terms) ready-made burrows, was home to some four hundred groundhogs. The burrows consisted of tunnels leading into underground tanks that were built into mounds of earth. From the inside these were ideal groundhog homes. What the animals did not realize was that each mound had a lid on it so that the researchers could reach them while they were hibernating.

Once the groundhogs were settled in for their winter hibernation, the scientists were able to open the lids of the burrows and pick up the tightly curled balls of fur. Unlike when they were awake (and to the great relief of the scientists), the groundhogs did not seem to notice. For the first time, the animals could even be described as cute. The researchers collected extracts of blood, fats, proteins and steroids. They measured, analysed and recorded. The evidence pointed to there being a chemical – some active substance – that let the groundhogs hibernate without coming to harm. All the research team had to do was find it.

But Bill Bigelow was not planning to wait until he had discovered the elusive secret of hibernation. His hypothermia research on dogs had already proved successful, and safe* enough to try on humans. Now the Canadian surgeon just had to wait for the right patient. However, if he had been hoping to make it into the history books, he was about to be beaten to it.

OPENING UP THE HEART

University Hospital, Minneapolis, 2 September 1952

The green-tiled operating room was the modern equivalent of an old Victorian operating theatre. Instead of a raked gallery surrounding the operating table, spectators could observe from a room above, through glass portholes in the domed ceiling. And today's operation would certainly be worth watching.

Some of the brightest, most ambitious and daring cardiac surgeons are working in Minneapolis. Today, F. John Lewis is leading the surgical team. He is assisted by a young surgeon called Walter Lillehei, a man who will come to epitomize the heart surgeon: confident, resilient and, it will later become apparent, something of a showman. Above all, these are men (and they are *all* men) who are not afraid to fail.

The patient is a thin, frail five-year-old girl named Jacqueline Johnson. She has been diagnosed as suffering from a hole between

* *Well, reasonably safe. Experiments had suggested that cooling the body too much could stop the heart altogether.*

the two upper chambers (the atria) of her heart. Without surgery she is unlikely to live much longer. Her heart is already swollen and she is becoming weaker by the day. The anaesthetist puts her to sleep (using a muscle relaxant to prevent her shivering) and the surgical team wraps a special blanket threaded with rubber tubes around her. They tie the sides of the blanket together with wide ribbons of cloth and turn on the taps to allow cold water to pass through the tubes.

It is a slow process to gradually cool the girl down – it takes twenty-five minutes before her temperature has dropped just one degree. Eventually, after two hours and fourteen minutes, her body's core temperature is down to 28°C – 9 degrees below normal. And as the girl's temperature falls, so does her heart rate. Jacqueline's heart is now beating at half the normal rate. According to calculations based on Bigelow's research, if surgeons usually had four minutes to operate on the heart to avoid starving the brain of oxygen, they now have six. But is six minutes enough to cut open the girl's heart, repair the defect and sew it back up again? Can those extra two minutes make the difference between failure and success?

The surgeons untie the blankets and Lewis cuts open Jacqueline's chest. Her heart is beating slowly as the surgeon prepares to clamp off the girl's circulation. Lillehei starts his stopwatch.

The six-minute countdown begins.

Lewis works slowly and precisely. Unnecessary haste could be fatal. He tightens tourniquets around the veins entering the heart and the arteries leaving it. The blood stops moving around Jacqueline's body, but her heart keeps beating. Lewis cuts into the right atrium to expose the inside of the heart. Unlike closed-heart operations, where surgeons operate in a river of blood, Jacqueline's heart is practically dry. Lewis can clearly see what he is doing. The defect is exactly as he had expected: a hole between the left and right atrium. He begins to sew.

Two minutes left.

Lewis finishes sewing and pours some saline solution into the heart to test the repair. There is a leak. He puts in another stitch and tries the saline again. The hole is closed.

One minute left.

Lewis starts to suture together the thick muscle of the heart wall. The muscle is still beating but the rhythm is becoming weaker, the beat irregular.

Thirty seconds.

Lewis releases the clamps across the arteries and veins. Blood begins flowing. The surgeon grasps the heart in his hands and begins squeezing to help it back into its natural rhythm.

Time up.

He closes the girl's chest as quickly as he can and carries her over to a bath of warm water (actually a watering trough ordered from a farm catalogue). Her heartbeat becomes stronger. She is going to be OK. Five-year-old Jacqueline Johnson leaves hospital eleven days later. She will grow up to have two children of her own. It was an incredible surgical advance: open-heart surgery had arrived.

MEANWHILE, BACK AT THE GROUNDHOG FARM

Although Bill Bigelow did not get to perform the first successful open-heart surgery on a human patient, he was undoubtedly pleased that his theory had been proved right. Many patients, particularly young children, would owe their lives to him. Hypothermia bought surgeons valuable extra minutes – enough time to carry out procedures that had previously been impossible. Bigelow continued to work to improve the techniques of cardiac surgery. He developed the first electronic pacemaker. He also continued to study the groundhogs.

Bigelow's team had been collecting groundhogs for almost ten years. The farm was thriving; the little bastards were still biting. Back in the lab the doctors were taking extracts from the animal's brown fat deposits – pads of fat that the researchers decided were the key to hibernation. These samples were analysed and their chemical composition checked. Finally, in December 1961, it appeared that all the research effort had paid off – one of the tests revealed a completely new substance. Could this be the mysterious chemical that allowed the groundhogs to hibernate?

A small amount of the substance was extracted and injected into some guinea pigs. The animals were then cooled down to low temperatures – much lower than they had previously been able to endure. There were no ill effects. This could finally be it. There was great excitement among the team. A vial of the new chemical was personally delivered to the National Research Council in Ottawa for further tests. It was even given a name: Hibernin.*

The hospital appointed the finest patent lawyer, and Bigelow filed a patent (no Minnesota surgeon was going to get his hands on the product of his research this time). NASA made enquiries – perhaps

* *Its full chemical name was 1-butyl, 2-butoxy-carbonyl-methyl-phthalate.*

they could use this substance for astronauts on long-duration space missions? A few journalists got wind that something was going on, but the researchers kept their silence. One of them even delayed a promising job offer so that he could spend more time with groundhogs.

They decided to try out Hibernin on patients. After all, the guinea pigs had survived. The surgeons operated on two people suffering from holes in the heart. The human guinea pigs were hooked up to some tubing to enable Hibernin to be injected. Bigelow found he could cool the patients down to around 18°C – four degrees lower than anything they had achieved before – which bought the surgeons much more time. Both operations were successful. The only peculiar thing they noticed happened after the operation: the patients were sleeping for much longer than usual. They seemed groggy. It was strange, but the nurses in the recovery room said it was almost as if they were drunk.

Now that Hibernin had been proved to work there was immense pressure to publish the results of the trial. Everyone was lined up for a major media event to make the big announcement. This would be a significant achievement for Canadian science. Then Bigelow received a letter from the patent office in Washington DC. The letter said that the chemical had already been patented. 'Hibernin' had been used for some twenty years as a plasticizer – employed to make intravenous plastic tubes pliable. Bigelow was incensed. How annoying that a biological extract from groundhogs turned out to be the same stuff as an industrial chemical. Still, the team had better do one final check as a scientific formality. Some clean plastic tubing was cut up and placed in water. A few hours later the scientists analysed the water. They extracted Hibernin.

Rather than having extracted a miracle substance from groundhog fat, the surgeons had simply flushed out the plasticizer from the tubing used in their research. The plasticizer was a potent form of alcohol. This explained why the patients acted as if they were drunk – they were. It says a lot about Bigelow and his management style that he and his team were able to laugh about it. Ten years of visiting the groundhog farm in the bitter winter. Ten years of groundhog bites. Ten years of hard research work. Thank heavens they had held back on the publication.

They never did find the secret of hibernation. The groundhog farm was shut down and the researchers moved on to other things. Bigelow would often cite the experience as a humbling example of 'intellectual humility'. However, the research did not go completely to waste. It opened up a whole new area of study into the use of alcohol in hypothermia. It turns out that alcohol dilates the blood vessels, allowing smoother cooling and causing less long-term damage. A refinement of Bigelow's hypothermia technique is still being used in operating theatres today.

DR LILLEHEI HAS AN INTERESTING PROPOSAL

'Breakthrough' is one of the most overused words in science and medicine. Most progress is incremental – small changes in procedures or techniques, refinements of treatments and technologies. However, when it came to surgery of the heart, the only way to progress was through daring new breakthroughs: Hill sewed up a wounded heart; Harken cut into a beating heart; now Lewis had performed the first successful open-heart surgery. These surgeons had the courage to try completely new ideas on real living patients.

Hypothermia was undoubtedly a major breakthrough, allowing surgeons genuinely to cure some of the worst heart defects. But hypothermia was severely limited by time. Doctors had only a few minutes to clamp off the heart, open it up, fix the defect, close the heart and restart the circulation. Hypothermia increased the time they had, but it could not stop the clock altogether. And when it came to open-heart surgery, there were so many things that could go wrong. There could be a problem with the anaesthetic, or a difficulty during the cooling of the patient. The surgeon might accidentally sew through a hidden nerve, interrupting the heartbeat or even stopping it altogether (this was known as 'heart-block').

What these pioneering surgeons feared most was a misdiagnosis. Jacqueline Johnson, the first open-heart patient, was suffering from an ASD (atrial septal defect). This was a hole between the upper two chambers of the heart, the atria. Patients could also have holes between the two ventricles – a much more serious condition – or worse. There could be defects in valves, in muscle or nerves. There were some heart defects that many surgeons thought might never be conquered, not least the sinister-sounding tetralogy of Fallot.* This disorder involves not only a hole between the ventricles, but obstructions between the right ventricle and the lungs, a thickening of the right side of the heart and a distortion in the aorta, the main artery from the heart.

* *The peculiar name for this congenital condition comes from Etienne-Louis Fallot, a Marseilles surgeon who first described it in 1884.*

In the 1950s surgeons had a limited number of tools at their disposal to work out what was wrong. They could X-ray the heart, listen to the heartbeat and study the rhythm on an electrocardiograph. Usually they got the diagnosis right, but sometimes it was wrong. They would open up the heart to find a larger hole than they expected, two holes instead of one, or multiple problems. And however quickly they worked and however brilliant their technique, the surgeons would run out of time. If that happened, the patients – invariably children – died on the operating table.

Hypothermia slowed the clock down, but surgeons now wanted to stop it altogether. To fix some of these more complex problems they needed to be able to isolate the heart completely. They needed some way of clamping off the circulation without jeopardizing the rest of the body by shutting off its blood (and hence oxygen) supply. Back at the University of Minnesota, surgeon, and now associate professor, Walter Lillehei had a brilliantly simple, if somewhat bizarre, idea. Why not keep the patient's blood circulating with someone else's heart?

The theory went like this: as well as the patient, a healthy person would be brought into the operating theatre. Arterial blood from the healthy person's body would be pumped across to the patient. This oxygenated blood would be passed directly into the patient's arteries to circulate around his body. Then, instead of returning the blood to the patient's heart, deoxygenated blood would be returned to the donor. The concept became known as cross-circulation and held enormous promise. During the period the two people were connected, their blood mingling together, surgeons would be free to open up the patient's heart. They then had plenty of time to fix any major defect.

The healthy donor would obviously need to have a matching blood group. But as the donor would usually be a close family member – ideally a parent – this would not be a problem. And what parent wouldn't do all they could to help their dying child? Yes, OK, it was risky taking a perfectly healthy adult into an operating theatre, sedating them and hooking them up to someone else, but wasn't it a risk worth taking? What could possibly go wrong?

Lillehei was not one to shy away from risk – particularly when the reward was so great. If this worked, he would be able to save any number of children from an early death. He began experimenting on animals to refine the technique. He acquired a pump – normally used in the dairy industry to move milk – and some plastic tubing designed for beer taps. He had to work out the layout of the operating theatre, the staffing and procedures for two patients. Above all, he had to make sure that the system he devised to connect the two patients was airtight. Any foaming from the pump, in fact just one tiny bubble, was enough to induce a stroke. It could leave the patient or the donor with permanent brain damage. If something went awry during the operation, one or both of them could be killed. Lillehei had invented the first surgical procedure with the potential for 200 per cent mortality.

SO MUCH FOR THE THEORY

University Hospital, Minneapolis, 31 August 1954

Howard Holtz was an ordinary man with an ordinary job. He spent most of his working life outside, maintaining the Minnesota highways. The twenty-nine-year-old was married with three perfectly healthy children, and another on the way. There was nothing particularly unusual about Howard Holtz. Except his blood. He had AB negative blood – the rarest of blood groups, found in only 1 per cent of the population. The blood was on a donor register. Of course, it's one thing to donate blood, quite another to donate your entire circulation, but this is precisely what Howard was asked to do when he was approached to act as a donor for one of Lillehei's operations.

Lillehei had been performing cross-circulation operations since March, and the procedure was reasonably well established, even though the risk was still considerable. The first operation on a sickly one-year-old baby boy had been successful. The patient and donor were connected for some nineteen minutes. Unfortunately, the boy died eleven days later from another complication.

In April the surgeon had operated on a three-year-old boy and four-year-old girl. Both operations were successful – successes that the proud showman Lillehei revealed to a press conference a week or so later. He even wheeled out the cute little girl so that she could be photographed with her parents, and her father, the donor, could be questioned by the pressmen. They were told how close the poor girl was to dying, the pneumonia she suffered from, and the oxygen tent she once had to live in. Now she could grow up to lead a normal and healthy life (and she would).

The papers described cross-circulation variously as 'miraculous', 'daring' or even 'impossible'. Some patients died, but those cases didn't get reported. There were no official mortality figures. In 1954 cross-circulation was the best chance many extremely sick children had of surviving into adulthood. Whether to go ahead with the operation was an awful decision for parents to make, but, after careful consideration, most decided it was their child's best hope. As Minneapolis was the only place in the world where this operation was being performed, most parents considered themselves lucky even to have the opportunity.

For Mike Shaw's parents, Lillehei offered the chance of a miracle cure. Ten-year-old Mike was seriously ill. He had been diagnosed as suffering from tetralogy of Fallot and had been in and out of hospital since birth. You could tell he was sick just by looking at him. The boy was thin and pale, his skin so tinged with blue that he was practically translucent. His ears stuck out from his wan face, giving him an emaciated appearance. Mike could walk only a few paces before becoming breathless. Without surgery he had only months to live. Lillehei might be able to save his life.

Mike's parents agreed that Lillehei should go ahead with a cross-circulation operation. They were aware of the risks, but trusted the surgeon, who was at least honest about their son's chances (although as this was the first attempt to correct tetralogy of Fallot, no one really knew for sure). The boy's blood was tested so that they could decide which family member would make the best donor, but then they hit the snag. With AB negative blood, neither Mike's parents, nor seemingly any other relatives, matched. Would a complete stranger be prepared to help?

When Lillehei explained the situation to Howard Holtz, the highway worker agreed to lend his body to the procedure. A complete stranger to Mike Shaw, Howard figured that if his own kids were sick, someone would do the same thing for him. A child's life was at stake, and Howard realized that a 'no' from him amounted to a death sentence for Mike. As far as the safety of the operation was concerned, none of the previous donors appeared to have suffered any ill effects. Any risks (and Lillehei had explained clearly that there were risks) were surely worth taking. Howard met Mike and the boy's family. The operation was scheduled.

Because cross-circulation involved two patients, it required two teams of surgeons. It is incredible that so many could fit into the small operating theatre. The room seems to be teeming with people, with little space between them. Everyone is gowned and masked, all slightly anxious. On the left lies the heart patient. At his head the anaesthetist and his assistant. They need to keep the patient's lungs replenished with air until the cross-circulation is connected; after that they are unusually powerless. A low curtain separates the anaesthetist from Lillehei and his assistants.

Even with his hat and mask on, it is easy to spot Lillehei. A long scar runs down his neck and disappears beneath his gown. The scar is evidence of major surgery to remove a tumour, and gives his head a lopsided appearance. Above the table a set of lights is angled downwards, but Lillehei also wears a head lamp on his forehead so that he can see clearly into the bloody hole in the boy's chest. The lamp, which is plugged into a socket in the floor, looks like it has been cobbled together from an old desk light, and becomes uncomfortably hot above the surgeon's face.

To the right of the main operating table lies Howard. He has also been put to sleep. This is not strictly necessary for the operation, but avoids any distress (or even boredom) on the donor's part. It is important to keep the anaesthetic as light as possible –

any drug circulating in Howard's body will also circulate in Mike's. The anaesthetist also makes sure the donor's breathing is regular. As long as the two patients are connected, Howard is breathing for two. A surgeon has made an incision in Howard's right leg (left as you look at him) and inserted a tube into his femoral artery. Another tube enters the main vein in the leg – the great saphenous vein.

Between the two operating tables snake the beer tubes full of blood. Brightly coloured oxygenated blood flows one way across the operating theatre and darker venous blood flows back the other. The blood passes through the dairy pumps to regulate the pressure and make sure the boy's fragile circulation is not overloaded. The pumps make a smooth, rhythmic sound as a line of small mechanical fingers press the blood along the tubes. Nurses move between the two patients, a surgeon monitors the flow of blood, Lillehei cuts away at Mike Shaw's heart.

For those observing through the windows of the gallery above – even the most experienced of surgeons – this is a remarkable operation to witness. Probably the most daring, ambitious and perhaps downright foolhardy they have ever seen undertaken. As Lillehei cuts and sews slowly, methodically beneath them, some of those watching are mentally calculating the odds on Mike Shaw and Howard Holtz both coming out of the operating theatre alive.

The pump is switched off. Mike's heart takes up the strain. Howard is disconnected. The donor leaves hospital after a few days. Not long after, so does Mike Shaw – he is cured. It is another miracle for Lillehei's revolutionary and 'impossible' surgery.

The operation had transformed Mike Shaw from a sickly patient to a healthy, active boy. At the time of writing, Mike and Howard are still very much alive. Mike grew up to become a musician – a bass guitarist – and, thanks to Lillehei's operation, has lived life to the full. At eighty-two, Howard is also fit and healthy, and regularly goes line dancing. Several years after leaving hospital, Mike's mother complained to Lillehei that her son was now playing in a band, staying out late at night and dating a lot of girls. Before the operation she had been worried that he couldn't do anything; now she worried he was doing too much!

Lillehei was a hero to the patients he saved, but unfortunately not every case was so successful. Later that year he had a series of failures – complications arose or a misplaced stitch resulted in heart-block. Sitting with parents telling them the worst news possible is something few surgeons get used to but, unlike some, rather than delegate the responsibility Lillehei made it his job to talk to the parents himself. Despite his self-belief and bravado, Lillehei shared their grief. But somehow he was able to recover, ready to attempt another operation the following day. At one point he was close to abandoning the procedure altogether, until persuaded by his boss to keep going. In the end he performed cross-circulation operations on a total of forty-five sick children. Twenty-eight survived surgery and most went on to lead normal, healthy lives.

It was the welfare of donors that finally brought an end to the operations. On 5 October 1954 Geraldine Thompson was hooked up to her daughter and the pump between them switched on. Lillehei began to operate, concentrating on the girl's heart. However, someone else in the operating theatre was not doing his job properly. A bubble of air had got into the system. The operation was halted, but it was too late. Mrs Thompson was left severely brain damaged. The observation that this was an operation with the potential for 200 per cent mortality had almost come true.*

No other surgeon in the world dared to attempt cross-circulation, although some tried ideas that seem even more absurd. At the Hospital for Sick Children in Toronto, surgeon William T. Mustard was experimenting with monkey lungs. Just before the operation, he would anaesthetize and kill several monkeys, remove their lungs and clean out the disembodied organs with antibiotics. The lungs would then be suspended in jars of pure oxygen and connected to the patient. During a series of operations on twenty-one children, only three of Mustard's patients survived.

Another surgeon and researcher, Gilbert Campbell, tried a similar experiment with the lungs of a dog. After a successful trial during a routine operation (not on the heart) he recruited Lillehei as lead surgeon to give it a go during a tetralogy of Fallot case. The patient died shortly after the operation, but later attempts were more successful, most famously in an operation on Calvin Richmond, a thirteen-year-old Afro-American boy from Arkansas who had been badly injured in a road accident and was seriously ill. Doctors concluded that he was suffering from a hole in the heart, but there was little they could do. His only hope lay in the miracle surgery being conducted by Walter Lillehei at University Hospital, Minneapolis.

* *Lillehei told Mr Thompson that the failure of the operation was due to an error. He clearly felt terrible about it and, as he held liability insurance, suggested that he sue on his wife's behalf for a reasonable amount. Unfortunately, once the lawyers got involved, a reasonable amount became millions of dollars, and the case ended up in court. The court ruled that Mrs Thompson had been fully aware of the risks, so the family ended up with nothing.*

A fund-raising campaign involving a Little Rock newspaper and TV station raised enough money to send Calvin to Minnesota for treatment and he was flown north courtesy of the Arkansas Air National Guard. However, on learning of cross-circulation, the boy's mother declined to participate in the operation. A volunteer was sought from the local prison instead. When none came forward – for fear of their 'white' blood and Calvin's 'black' blood mixing – Lillehei decided to use the dog-lung method. The operation went without a hitch, the animal lung oxygenated Calvin's blood while the boy's damaged heart was repaired. The success was widely reported, although most correspondents skated over the bit about the lungs from the dead dog.

If cross-circulation had its faults – and its potential to leave both participants dead was a downright terrifying one – then employing monkey and dog lungs was hardly any better. Lillehei used the dog-lung technique a few times more, but concluded that it was far from ideal. At least Lillehei knew when to stop. As for the monkey lungs, you can only have great sympathy for the desperate parents who put their trust in William T. Mustard. Something better was needed. And while the surgeons in Minneapolis were using other humans or animals to oxygenate the blood, surgeons elsewhere were turning to machines.

DR GIBBON'S REMARKABLE INVENTION

Philadelphia Jefferson Hospital, 6 May 1953

The operation was going well. Eighteen-year-old Cecelia Bavolek lay on the table, her chest cut open to expose her beating heart.

Dr John H. Gibbon Jr was relieved that the diagnosis had proved correct – Cecelia was suffering from an atrial septal defect – a hole in the heart between the two atria. His blood-splattered hands began to stitch the two sides of the one-inch hole together. For Gibbon this was a well-practised procedure, though all his previous successes had been on cats and dogs. His first, and until now only, attempt on a human patient had ended in death on the operating table.

Gibbon worked slowly, methodically and precisely. As usual, the operating theatre was crowded. There were other surgeons huddled around the table, plus assistants and scrub nurses to pass instruments. The anaesthetist monitored the girl's blood pressure; an assistant passed the surgeon some scissors. Gibbon was not relying on hypothermia to cool his patient; neither was he using cross-circulation or some other animal's lungs to pump and oxygenate the girl's blood. He was trying out the latest version of his great invention – the heart-lung machine – which was gurgling, humming and clunking beside him.

Gibbon's heart-lung machine looked (and sounded) like something out of a 1950s B movie, where the unhinged scientist meddles with forces he doesn't fully understand. But, on the face of it, there was nothing even slightly eccentric about Gibbon. He had a reputation for calm professionalism; he was well respected by his colleagues and, remarkably for a heart surgeon, was shy and self-effacing. Colleagues described him as a 'perfect gentleman', kind and considerate. If there was anything eccentric about Gibbon, it was his obsession with developing a machine to keep a human being alive during major surgery.

Gibbon had been working on the project since the 1930s. The early attempts were crude mechanical affairs, the size of a grand piano. Visitors invited to his lab to see the machine in action were issued with wellington boots. The giant machine needed buckets of blood to get it started, but once under way could sustain the life of a very small cat. Pretty soon the visitors would notice that the floor was getting wet and that they were walking around in blood. 'Uh oh,' said Gibbon, as pints of cats' blood sloshed across the floor. 'We've got a leak again this morning.'

Emulating the human heart and lungs within a machine proved to be a tough challenge. Replacing the heart itself was relatively simple: this could be done with a pump. As long as the circuit had some pressure controls and there was a safeguard against air getting into the system, an artificial heart pump could employ off-the-shelf technology (such as the dairy pump Lillehei used for his cross-circulation operations). The problem was the lungs.

Human lungs consist of a branched network of tubes, where gases are exchanged between the air and the blood. Oxygen from the air passes into the blood, and carbon dioxide passes from the blood into the air. The total surface area available for this exchange is an astounding 84 square yards – about the same area as a tennis court. Any machine needed either to include a similarly massive surface area (much bigger than your average operating theatre) or find some other way of getting oxygen into the blood. The obvious way was to bubble the oxygen into the liquid, but this was fraught with difficulties. If even the slightest single tiny bubble remained and was allowed to pass back into the patient's bloodstream, it could kill them. Gibbon favoured pumping the blood over a flat surface –

a plate or screen – to expose a film of blood to oxygen. As long as he could keep the blood flowing, this method seemed to work. The trouble came when the blood started to clot.

Over the years Gibbon's heart-lung machine became more refined. After the war the International Business Machine Corporation (IBM) offered its support and an engineer. Electronics were introduced to control the flow of blood and monitor the pressure and oxygenation process. The experimental animals got bigger and bigger, while the machine became smaller and more efficient. Even so, the heart-lung machine was still bulky and incredibly complex. Around the size and shape of two large top-loading washing machines bolted together, the contraption was so big that when it arrived at the hospital it had to be winched in through a window. But with IBM's help, it no longer resembled a crude Heath Robinson affair. Now it looked more like cutting-edge technology.

The machine was covered in switches, pipes and dials. Dials to measure acidity and pressure; electronics to monitor and control the flow of blood; even a back-up battery should there be a power failure. The top was a mass of plastic tubing, the sides hung with glass bottles. Rising from the upper surface was a rack of screens down which the film of blood would cascade to be exposed to oxygen. Snaking from it were two tubes – an input tube that would take blood from the patient's veins, and an output tube that would return oxygenated blood to the patient's body. Once it was hooked up, the machine would take the place of the patient's heart and lungs.

It is twenty-six minutes into the operation and Cecelia Bavolek is doing well. Blood that would normally pass through her heart and lungs is being diverted into the machine. It is being oxygenated and returned to her body. But something has gone wrong. The blood on the oxygenator screens is no longer running freely. It has started to clot. The pumps keep working and the pressure in the machine starts to build. On the operating table Cecelia is no longer receiving enough oxygen. The machine begins to foam. It is going to explode.

Vic Greco is responsible for the machine.* He has been working in the research lab with Gibbon and guesses what has gone wrong. Before it is hooked up to the patient the machine has to be 'primed' with blood. When they had done this earlier in the day, they had probably not added enough of the blood-thinning chemical heparin. But there is no time to analyse why it has gone wrong. They have to solve what is rapidly turning into a very messy crisis. Unless they can fix the machine, Cecelia Bavolek is going to die.

At the operating table Gibbon tries his best not to get too distracted. He works as quickly as he can, but the foaming is getting worse. The blood is beginning to back up around Cecelia's body, and her circulation is coming to a halt. Greco climbs up a stepladder to hold down the lid of the oxygenator to prevent Cecelia's blood from spraying around the room. Then Bernard Miller, who has been intimately involved in the technical development of the machine, starts rerouting the pipes. He figures that the only chance they have is to bypass the now useless oxygenating screens and turn the heart-lung machine into just a heart machine. This will at least get the blood moving and restart Cecelia's circulation.

* *The machine was usually the responsibility of Jo-Anne Corothers, but it had been decided that it would be 'better for the historical record' if a doctor ran the machine that day.*

The blood starts to flow again, only this time it is not getting any oxygen. This is circulation of sorts, but what hope does Cecelia have without any means of getting oxygen into her system? Gibbon carries on anyway. Cecelia's heart loses its rhythm and goes into fibrillation. Gibbon begins to stitch together the incision he has made. He uses an electric shock to get her heart beating again and it goes into a normal rhythm. She could yet live. At least nothing else can go wrong. But it isn't Gibbon's day. As the surgeon continues to work, Cecelia starts to come round from the anaesthetic. She struggles on the operating table. Gibbon closes her chest and puts the final stitches in her skin. Remarkably, her heart continues to beat; her breathing is normal. Within a fortnight, Cecelia Bavolek is discharged from hospital, the hole in her heart successfully closed.

The operation had lasted forty-five minutes. For twenty-six of those minutes her life had been sustained by a machine. It was proclaimed an 'historic operation'; the twenty-six minutes 'the most significant in the history of surgery'. Gibbon shied away from the publicity the operation generated, shunned the press and only grudgingly gave a few quotes to *Time* magazine (although he declined to be photographed with the machine). As far as Gibbon was concerned, the successful operation was the result of more than twenty years of research, and he had proved that a heart-lung machine could work.

Nevertheless, Cecelia's operation was a close-run thing – she was lucky to be alive. Just how lucky would soon become clear. Gibbon attempted two more operations using the heart-lung machine. Both operations were carried out on five-year-old girls. Each of them died in the operating theatre. Gibbon had had enough. He did not have the resilience of some of his colleagues to carry on regardless. Three of his four patients had died while connected to the machine. The surgeon decided not to operate with his machine again, and ordered a year-long moratorium on its use. Gibbon never returned to cardiac surgery.

But others believed that Gibbon was on to something. At the Mayo Clinic in Rochester, less than an hour and a half's drive away from where Lillehei was developing cross-circulation, another surgeon began work on refining Gibbon's machine. After two years of research, on 22 March 1955, John W. Kirklin was ready to operate. He decided to test the machine on eight patients – no more, no less. During the first operation on a five-year-old girl, the machine practically exploded. There was blood everywhere, but the patient survived. By May Kirklin had operated on his target of eight patients. Four survived. The odds were improving, although patients still had only a fifty-fifty chance of coming out of the operating theatre alive.

DR LILLEHEI RISES TO THE CHALLENGE

Minneapolis, 1955

Back in Minneapolis, Walter Lillehei was also working on a heart-lung machine, only his was a good deal simpler. Lillehei decided to try just the sort of system that everyone else had warned against – one that bubbled oxygen into the blood. He assigned the task of designing the new machine to Dick DeWall, a young doctor who had come to Lillehei with a design for an artificial heart valve – something DeWall had been working on in the evenings at home (as you do).

Lillehei decided not to mention to DeWall that everyone else believed that a 'bubble oxygenator' was impossible, if not downright dangerous. But then they had said the same about cross-circulation.

Bubbling oxygen into the blood is relatively easy. The problem comes with getting the bubbles out again. DeWall set to work with a couple of pumps (the same sort of dairy pumps that were being used for cross-circulation) and some plastic tubing, all held together with a few bits of tape and some metal hose-clips. The resulting machine looked too simple to be effective, but that simplicity was the beauty of the system. The blood from the patient's veins was pumped into a mixing chamber, where oxygen was bubbled through a large rubber stopper with hypodermic needles sticking out of it. The newly oxygenated bright red blood then passed through what DeWall termed a 'de-bubbler tube' – a diagonal piece of pipe filled with an anti-foam chemical to break up the surface of any bubbles. It was the same chemical used in factories to make mayonnaise. Finally, the blood flowed down a helical spiral. This was probably the cleverest bit and was designed to defeat any lingering bubbles. The heavier blood that was free of bubbles rolled downwards with gravity, while the lighter blood, containing bubbles of air, was forced back to the top. Finally, the blood flowed out of the helix through another dairy pump and back into the patient's arteries. The whole arrangement of pumps, bottles and tubes sat on a trolley beside the operating table. When the operation was finished the plastic tubes could be thrown away – no need for the complicated cleaning or difficult preparation of the Gibbon machine.

The Lillehei-DeWall bubble oxygenator was first put to the test on 13 May 1955. Unfortunately, the patient later died, but this did not appear to be down to any fault in the machine. There were so many other things that could go wrong with this pioneering surgery. By December one hundred operations had been performed using the machine. Most of the patients survived. The odds of open-heart surgery were improving. The machine was refined, improved and commercialized. Soon any hospital in the world could purchase one.

With the heart-lung machine, surgeons were able to operate on an open heart free of blood. They could take their time and see exactly what they were doing. However, one last problem remained: even with the heart-lung machine connected, the patient's heart kept beating. Placing precise stitches into a beating heart was difficult, and the slightest slip could end in disaster. What surgeons needed was some way to stop the heart beating altogether. Of course, the other thing surgeons had to be able to do was start it up again.

The answer came from a British surgeon, Denis Melrose,* who published his research in the *Lancet* in 1955. He devised an injection of potassium citrate. He later changed this to potassium chloride, a compound that disrupts the electrical signals in the heart. It's the same chemical that forms the basis of the 'lethal injection' used to administer the death penalty in some US states. When it was first tried out in Britain on a patient at Hammersmith Hospital in London, they had to consult a coroner and church leaders because technically, for the duration of the operation, the patient was dead. As for starting the heart again, this was done with electricity applied directly to the heart muscle.

* *Melrose was a remarkable surgeon. As well as his work on stopping the heart, he also developed a heart-lung machine, which was adopted by hospitals all over the world.*

Within the space of a few years, cardiac surgery had been transformed. From Harken's first quick incision into a beating human heart to remove a bullet, surgeons such as Melrose and Lillehei were stopping hearts altogether to open them up and correct major defects. With the ability to stop the heart came even more daring and intricate procedures. Patches were stitched across large holes; artificial heart valves were grafted on; arteries were replaced with synthetic tubing. Every year more and more patients were undergoing open-heart surgery, and every year more and more were surviving.

In 1958 the personable young Melrose made history by conducting open-heart surgery live on Californian television. The broadcast started at 7 p.m. when Melrose began operating on 'Tommy', the seven-year-old son of an American war veteran. For over four hours viewers were glued to their small black and white TVs as Melrose cut open the boy's heart. It was the highest-rated programme that night – real life drama with the genuine risk that the boy might die on live television. Tommy survived, but *his* heart was repairable. What about those hearts that were so badly damaged that no amount of surgery could fix them? Could the heart – the centre of the soul, the very core of the body – be replaced with another one?

THE NIGHT OF THE PIGS

National Heart Hospital, London, 1969

It was a desperate, last-ditch attempt to save a life. An experimental procedure to keep a dying patient alive.

One of the UK's leading cardiac surgeons, Donald Longmore, had put in the call. The farmer assured him that the pigs were on their way. He would deliver them himself in his Land Rover. It wouldn't take long. Everything else was ready.

In the operating theatre the male patient lay on the operating table, his chest open and tubes snaking across to the bulky heart-lung machine. Dark red blood flowed one way, bright crimson blood flowed back. The machine's regular beating rhythm was keeping the man alive. The anaesthetist, sitting beside a complex rack of gas canisters, calmly monitored the patient. A nurse placed some freshly sterilized instruments on the trolley; another kept an eye on the machine. All the lights and dials seemed to be indicating everything was OK. There was little else for the surgeons to do than wait. For the pigs.

They had decided to call this the 'piggyback' operation. The surgeons' plan was to graft a pig's heart and lungs into a patient so that the animal's organs would help keep the man alive. The operation had been conceived to help someone with serious heart disease. The pig's heart and lungs would work – or piggyback – alongside the patient's own heart and lungs to relieve some of the strain. At the very least it might keep this seriously ill man going for a few more months before the heart transplants pioneered in 1967 and 1968 were perfected and a suitable donor found. It might last even longer. It could even be another 'miracle breakthrough' the newspapers were so fond of reporting. As usual, the procedure had been tried on other animals and it seemed to work. The patient was seriously ill, his heart ailing. This experimental operation offered the only chance of survival.

Everyone waited for the pigs.

The farmer pulled up to the mews at the back of the hospital.

The first inkling Longmore had that it was going to be a long night was when Thompson, the head porter, called him. 'Mr Longmore, is that pig in a Land Rover in the mews anything to do with you?'

'Yes, it is.'

'Well, it has just got out and turned left along Wimpole Street.'

Reluctant to make its own valuable contribution to medical progress, the pig had escaped. It is surprising how fast a pig can run, especially when its life is at stake. Still dressed in their operating theatre gowns, caps, masks and boots, the entire surgical team gave chase.

The pig ran as fast as its little legs could carry it, but was no match for London's finest heart surgeons, who eventually caught it halfway up the road. The pig squealed in protest, but Longmore herded it back towards the hospital. It was five o'clock in the evening and people were heading home from work, so the street was relatively busy. Most passers-by paid little attention to the odd group in the road. Only one gentleman seemed a little perturbed. Raising his bowler hat, he said, 'Excuse me, sir. You are going the wrong way along a one-way street.'

Back in the operating theatre, the anaesthetized patient lay on the table. The heart-lung machine pumped and breathed on his behalf. The nurses and surgeons stood around. The clock ticked. Where was the pig?

The pig was with Longmore in the lift. He wasn't going to let it get away this time. There were also a few hospital visitors in the lift, but no, they didn't mind if the lift went straight to the top floor marked 'mortuary'. What business was it of theirs if the surgeon fancied having pork for his supper?

Arriving at the mortuary, Longmore had arranged for an anaesthetist to put the pig to sleep so that it could be killed and its organs removed. When the anaesthetist assigned to the task showed up, he turned out to be Jewish. He refused to kill the pig. Another anaesthetist was found, but by now Longmore was beginning to wonder if all this grief was going to be worth it.

In the operating theatre, the heart-lung machine continued to pump. The surgeons and nurses waited.

The heart and lungs were eventually removed from the pig, but now there was another problem: the patient was also Jewish. What were the chances? The patient himself was in no position to reassess the merits of the operation, so rather than panic (or pray), Longmore did the next best thing – he rang a rabbi.

When Longmore explained what they were trying to do, the rabbi went very quiet. The surgeon apologized for putting him in such a difficult position and understood if he didn't want to get involved. There was another long, somewhat muffled silence. Finally, the rabbi could hold back no longer. 'Sorry,' he said. 'I was trying to stop laughing.' The rabbi told Longmore that if this was a genuine attempt to save the man's life, then certainly he should go ahead. First the escaping pig, then the Jewish anaesthetist, now this. Another obstacle overcome. At last the surgeons could get on with the operation.

It was a relief to return to the operating theatre. Once he had changed and scrubbed, Longmore was ready to begin. The heart-lung machine continued to pump. The dark red blood flowed in; the bright red blood flowed out. The patient was still alive, the pig heart was ready. The operation could get under way.

The operation itself seems to be going surprisingly well. The heart is stitched into the patient's circulation ready to help keep him alive. The final part of the procedure involves a simple injection of calcium to the pig's heart. In humans calcium is used to increase muscle strength. However, as Longmore now discovers, it has a different effect altogether in pigs. The pig's heart sets like a stone. It is useless, and after all that effort the operation fails and the patient dies. It is little consolation to the surgeons that he would have died anyway, but at least they have learnt from the experience.

The story became known as the 'night of the pigs', and to cap it all, the feared hospital matron had been woken up by pig squeals and was furious. A member of the surgical team sent her pork chops for breakfast, which hardly helped.

FRAGILE HEARTS, FRAGILE EGOS

Groote Schuur Hospital, Cape Town, 3 December 1967

Compared to bullet wounds, hibernating groundhogs, cross-circulation and porcine predicaments, heart transplantation itself is a relative anticlimax. In 1967 the race was on to be the first surgeon to transplant a human heart. There was no knowing who would get there first, and although many surgeons talked of sharing their results and cooperating with their rivals, secretly most of them would admit that they wanted to be the one to get the credit and possibly a little bit of glory.

Many believed the first would be Norman Shumway, a surgeon at Stanford in California, who had been working for almost ten years on perfecting the heart transplant technique in animals. Shumway had presented his first heart transplant results on dogs in 1961, and was now ready to try it on humans. He had developed new combinations of drugs to prevent the heart from being rejected by the body (see Chapter 3). He even went on to suggest that transplanting a heart should be a relatively straightforward surgical procedure, and less likely to end in rejection than, say, a skin or kidney transplant.

Elsewhere in the United States, down in Mississippi, James Hardy was waiting for a human donor for a terminally ill heart patient. With no suitable transplant available, he tried transplanting the heart of a chimpanzee, but this proved unsuccessful. Meanwhile, many European surgeons were keen to steal a march on the Americans – there was national, as well as personal, pride at stake. So far, most of the cardiac 'firsts' had been by Americans. In France, hospitals were said to be ready, and in London Donald Longmore at the National Heart Hospital was waiting for the right combination of patient and donor, having spent the previous few months battling with a bunch of 'oily' bureaucrats from the Department of Health.

They were all beaten to it by a relatively unknown, although hardly unqualified, South African surgeon. Christiaan Barnard had been trained by Walter Lillehei in Minneapolis and had a good, if inconspicuous, surgical pedigree. His record for complicated open-heart surgery was remarkable. The chances of surviving a Barnard operation were extremely good. Like many of the most successful heart surgeons, he thought of the organ as merely a 'primitive pump' – one that demanded respect, but commanded no great mystical power, no soul.

Barnard had been studying the problems of heart transplants for some time, although much of what he learnt was from other surgeons rather than his own experiments. He travelled to see Shumway at Stanford, and visited Longmore in London to witness his heart-lung transplants on dogs. Barnard was personable, anti-apartheid and had something of a reputation – the handsome young man's interest in affairs of the heart extended beyond pure technical interest. While in London, he also attempted to seduce one of Longmore's nurses (with some success apparently).

Barnard had learnt from the greats, and in an obscure South African hospital, undisturbed by the authorities, he decided the time was right to put his knowledge to the test. On 3 December 1967 he removed the beating heart of a twenty-five-year-old female car crash victim, pronounced dead by a neurosurgeon. The recipient of this fresh young heart was lying ready in an adjacent operating theatre. Fifty-three-year-old diabetic Louis Washkansky had already suffered several heart attacks and, quite frankly, did not have long to live. The operation took two hours. Washkansky's new heart started beating and kept beating. A day later he was awake and talking. A few days later he was out of bed. Eighteen days later he was dead.

Washkansky died of pneumonia. The drugs used to suppress his immune system to prevent the heart being rejected had also left him open to infection. But no one remembers Washkansky. Christiaan Barnard was the hero of the day – an instant celebrity. Barnard would be received by the pope, entertained by presidents and prime ministers. He became the world's most eligible bachelor, dating a string of beautiful and famous women.

Some of Barnard's rivals expressed more bitterness than others. Longmore was pleased for him. Others muttered that they had done all the work only for Barnard to take the glory. And what glory. Who cared about the second man to fly across the Atlantic (Bert Hinkler), to run the four-minute mile (John Landy) or to climb Mount Everest (a matter of some debate)? Christiaan Barnard would be the man in the history books.

But while Barnard had won the main prize, there was still a degree of national pride at stake. If South Africa could do it, why not the United States, Great Britain or France? The same bureaucrats who had been so reluctant for Longmore's London hospital to carry out a heart transplant were now asking what he was waiting for. In January 1968 the second heart transplant was performed by Adrian Kantrowitz in Brooklyn, so Shumway – the surgeon who had spent so long developing heart transplant techniques – didn't even get that honour. Shumway's operation was the fourth, and by the time he came to operate later that month, Barnard had already performed a second heart transplant.

The first British heart transplant (the world's tenth) took place on 3 May 1968. The surgeon was Donald Ross (also a South African). Longmore's role was to collect and deliver the donor heart. For some reason this required a police escort through the streets of London. In fact, the whole affair became a major public event, with a large crowd of spectators, reporters and photographers gathered around the door of the National Heart Hospital. It was, of course, a great national achievement of a proud nation, etc., etc. However, the patient, Frederick West, died of an 'overwhelming infection' forty-six days later. And that was the problem: while the surgeons were getting the glory, none of their patients were lasting very long. In the first few years of heart transplant surgery, patients survived on average just twenty-nine days. Despite all the euphoria, the awful truth was that heart transplants were difficult, dangerous and complicated.

There are few surgeons as well known as the pioneers of heart surgery. Heart surgeons were courageous, daring and bold. Heart surgeons stood apart from the rest, almost every operation a matter of life or death. When they succeeded, they saved lives. When they failed, they had to be prepared to come back the next day and try again. Many of them had personalities to match their abilities. Some were self-confident, others were egotistical or arrogant. A few were foolhardy or seemingly oblivious to risk. Most heart surgeons were revered by their patients; many became national or international celebrities – household names courted by the media, their faces on the front page of *Time* magazine. Few people could name one of today's heart surgeons, but then, thanks to pioneers such as Harken, Bigelow, Lillehei,* Gibbon, Melrose and Barnard, major open-heart surgery has finally become routine.

* *There is a curious footnote to Lillehei's career. In 1973 he was found guilty of tax evasion. Although he was undoubtedly at fault, his crime was more one of carelessness than deliberate evasion. He had always been bad at keeping financial records, and had performed many operations for free. He carried out his last operation in 1973, but was to maintain a keen interest in heart surgery until his death. At Lillehei's eightieth birthday party in 1998 many of those invited to celebrate owed their lives to his expertise. He died a few months later, but a great many of his patients live on.*

CHAPTER 3
DEAD MAN'S HAND

THE STORY OF A TRANSPLANT

March 2000

From a distance, there was something odd about New Zealander Clint Hallam. When he walked towards you it became obvious that one arm was longer than the other. Up close the arm was even more disturbing, verging on the grotesque. Anyone who saw his hand would remember it for ever, perhaps in their nightmares.

Hallam recalls sitting in an aircraft next to a nice old lady. They got talking. The lady recognized Hallam from somewhere, but couldn't quite place him. Then she happened to glance down at his right hand. Recoiling in shock, she pressed the call button to summon the flight attendant and asked to be moved to another seat. The lady turned to Hallam to apologize. She said she had nothing against him personally, only she couldn't bear sitting next to someone who was wearing a dead man's hand.

One of Hallam's close friends made a similar confession. Hallam could never understand why, when they met, his friend would always grasp Hallam's wrist rather than shake his hand. Hallam reckoned it was an act of kindness, to avoid any risk of injury to his limb. When he asked about it, his friend confessed to finding Hallam's hand quite horrific. He wasn't the only one. The recipient of the world's first hand transplant was beginning to realize that many people found his new hand repulsive. Now, eighteen months after the operation, even Hallam was beginning to have some doubts.

From the shoulder down, the upper part of Hallam's right arm was perfectly normal. Its skin tone matched Hallam's dark complexion and was covered in black hairs and dotted with freckles. Then, just beyond the joint of the elbow, there was a sharp division where the skin became pale and practically hairless. It was as if Hallam was wearing a long white glove. There was a bulge – a swelling – where the brown skin met the white. It was a long white glove that didn't quite fit.

The underside of Hallam's lower right arm was bruised and damaged. The skin was inflamed, raw and angry – as if he had been burnt. Beyond the wrist the hands were similarly swollen. The skin was peeling; there were ulcers and the flesh was shiny. It looked like the outer layer of skin had been stripped away. When it came to the fingers, the decay was even more pronounced. The fingertips were crusty and sore, the yellow nails gradually separating from the flaking skin underneath. Small wonder the woman on the plane chose to move to another seat.

Hallam had received his hand transplant on 23 September 1998 at the Edouard Herriot Hospital in Lyon, France. The operation took almost fourteen hours and was a brilliant technical achievement. Earl Owen led a team of some the world's most experienced transplant specialists from Australia, Britain, France and Italy. France was chosen to host the operation because of its laws on organ donation. There you have to opt out of donating your body to medicine, rather than opting in. As a result, almost everyone who dies becomes a potential donor, so many more donors are available – perfect if you are waiting for a new arm and hand.

The donated limb came from that all too frequent source of body parts – a motorcyclist. The limb was matched for blood type and tissue type, but, as it turned out, not appearance.

The surgical technique used to join together Hallam's stump and the dead motorcyclist's lower arm is known as microsurgery and is fantastically intricate. The surgeons wore powerful magnifying lenses and employed precise instruments, tiny needles and the finest of threads.

First they joined together the two bones of the forearm to hold the limb stable. Then they connected the blood supply – the arteries and veins – to keep the tissue alive. Once the blood was flowing, they stitched together the muscles and tendons and reconnected the nerves. Finally, the surgeons were able to join together the skin.

At the inevitable press conference held shortly after the operation, Owen described himself as 'very happy'. It was a moment of surgical glory. 'We all have big smiles,' he said, and gave the operation a fifty-fifty chance of long-term success. Others were equally enthusiastic. Eminent British transplant surgeon Nadey Hakim called it an 'incredibly exciting breakthrough,' adding, 'to see a man restored with an arm is tremendously satisfying'.

The new arm had to be kept immobile for a few weeks while the graft healed, but everyone was optimistic that the patient would develop the full use of his limb. Hallam himself was overjoyed. It was incredible to see fingertips at the end of his arm again. With a new hand, he had been given a new life. It was a surgical miracle.

Hallam had spent years waiting for the operation. He lost his original hand in an accident with a circular saw in 1984 while he was serving time at a prison in New Zealand.* His severed hand had been sewn back on, and although it looked OK, it had little function and was all but useless. A few years later Hallam decided to get rid of the hand altogether and opted instead for a prosthetic limb. This hadn't worked out either, and he told the BBC that he had never been able to accept having a lump of plastic attached to his arm. It was not natural. Perhaps one day the technology would be available for a hand transplant.

However, a few months after the 1998 operation Hallam was struggling to overcome his disappointment. It wasn't just the obvious mismatch between his new arm and the old one, so much as the practicalities. The new arm did not work very well. There was only limited movement: Hallam could move the limb and bend his new fingers to a limited extent, but he said he was almost more crippled with the hand than he had been with a stump.

There was a marked contrast between Hallam's experiences and

* *The circumstances of the injury remain somewhat vague. The surgeons who carried out the 1998 transplant operation were unaware that Hallam had sustained the injury in prison. Hallam's somewhat chequered past was revealed by the media following the operation. This did little to endear Hallam to the surgical team. Owen said later that although they had conducted psychiatric tests, they should have looked more closely into Hallam's background.*

the surgeon's rhetoric. They were claiming the operation as a great success and told the media how Hallam could grab things, pick up a glass and even write with a pen. They were also pleased that he could feel pain and temperature on both sides of the new grafted hand. But they weren't the ones who had to deal with the side effects.

To avoid the transplanted limb being rejected by his body's immune system, Hallam had to take a cocktail of different drugs. There could be eleven tablets to swallow in the morning, four at lunchtime and eleven more in the evening. The exact amount varied from week to week, but Hallam was usually being prescribed some combination of steroids, anti-rejection tablets and immunosuppressants – drugs that, as the name suggests, suppressed his immune system. He was also taking pills to help his fingernails recover. Every day he had to count out the various tablets to make sure he didn't take too many or too few. The pills were keeping his arm alive, but they were also having other effects. Hallam had started to develop diabetes and needed to take insulin to control his blood sugar. Physically, he was also changing. Hallam was used to keeping fit and had been in reasonably good shape. Now he found he was growing breasts. Worse, the powerful drugs increased his risk of developing cancer.

And it was not just the physical limitations that were taking their toll. Hallam was beginning to realize that there was a psychological price to pay for having a dead man's hand attached to his body. Aside from the mental anguish for any man of growing breasts, the transplant increasingly looked and felt like it did not belong. Other people would comment, saying how white the transplant looked, or how the new hand was smaller than the other one. Hallam realized how angry he was with the doctors for not waiting for a hand that was better matched. It was as if, he said, they were more interested in the transplant than the person. He had dreamt for years of having a new hand, but the reality was proving increasingly uncomfortable. What happened next became inevitable.

By 2001, Hallam had begun to take fewer and fewer of his prescribed drugs. Every time he got sick his immune system struggled to cope, so he decided the answer was to cut down on the immuno-suppressants, but the effect on his hand was hideous. Attacked by his own body's immune system, the limb had all but died; the flesh was rotting away on the end of Hallam's arm. He had lost all feeling; it was a wonder the infection hadn't spread to other parts of his body. Hallam told the BBC he felt 'mentally detached' from the hand. He had had enough and begged the doctors to remove it.

In February 2001 Clint Hallam's hand was amputated by Nadey Hakim, one of the surgeons who had helped attach it in the first place. The procedure, at a private London clinic, took ninety minutes. Hallam was relieved to see it gone. Some of the surgeons who had enjoyed such acclaim only a few years before were angry that their work had been in vain; that their patient had not persisted with his medication.

In retrospect, Clint Hallam was probably the wrong person to receive the world's first hand transplant. If he had done exactly what he was told – had taken his medication, had followed doctor's orders – he might still be walking around with a dead man's hand. Even so, he would have had to cope with the physical symptoms, the strict drug regime and probably a shorter lifespan as a result of diabetes or perhaps cancer. Eight years after his transplant, Hallam was once again fitted with an artificial limb. He claims he has no regrets about having the transplant; his only regret is being the first.

The case of Clint Hallam illustrates the barriers that have to be overcome for successful transplantation – whether it is the transplantation of a hand, finger, kidney or heart. The first barrier is simply one of technique. It has taken more than a century to develop the surgery of transplantation. Stitching together blood vessels is difficult enough, let alone trying to join muscles, tendons and nerves. But the operation itself is only the start. Next there is the enormous problem of rejection. The body's immune system will fight any alien tissue. Even with the latest drugs, rejection is still a major hurdle for transplant surgeons.

The final problem is more subtle, and is downplayed or ignored by surgeons at their peril. Any transplantation involves overcoming a psychological barrier. What is the effect of having a dead man's hand transplanted on to your arm? It certainly bothered that woman on the plane. Did Hallam ever think about the motorcyclist who had died? Before the transplant that hand had been gripping a motorcycle handle – an essential part of a whole different human being. What about the psychological effects of a kidney or heart transplant or even a face?

The surgeons who had operated on Clint Hallam had overcome the technical problems. Until he stopped taking them, the drugs had countered his body's physical rejection. But the surgeons had failed to overcome the final barrier. Consciously or not, it was Hallam himself who had rejected the hand.

THE DEADLY DENTIST

James Spence (and Sons), Soho, London, 1765

James Spence was always very discreet. No one, he assured the young lady, need ever know that she had visited him. Of course, nowadays he rarely conducted these procedures himself. He left most of the day-to-day work to his sons. However, for this fine lady he would make an exception. She was fearful of going to see anyone else (there were so many charlatans about these days). Spence had already established the finest reputation in London for pulling teeth, and was proud to call himself a dentist, even though 'dentistry' was only just starting to establish itself as a respectable profession. There was no one better to go to for a tooth transplant.

Spence preferred to use living donors for his tooth transplants. They were easy to come by and it avoided the repulsion many people felt at the idea of eating food with teeth from the dead. Mind you, teeth taken from cadavers were a lot cheaper, and many dentists did a roaring trade in teeth extracted from the mouths of soldiers killed on the battlefield.

Nevertheless, today Spence needed teeth from the mouths of young women. Earlier that morning he had dispatched a servant to locate suitable donors in the neighbourhood – women who would be willing to give up their front teeth. They would be handsomely rewarded (well, it would be handsome to them; the expense would make only a small dent in Spence's substantial profit margin). By mid-morning, several young women were queueing in the alley behind Spence's offices. He planned to take a few teeth, maybe a couple from each woman, to see which ones produced the best fit.

His patient arrived accompanied by a friend for support. Spence ushered the women into his consulting room. The patient sat down on one of the plush, high-backed leather chairs. At first glance she was something of a beauty and would, he thought, have no shortage of suitors. But when he took a look at her mouth he realized it was little wonder she had come to him for help. Her teeth were in a terrible state. Her mouth stank of decay, with black rotten stumps emerging from raw, inflamed gums. She was worried about the pain she was going to experience. Spence reassured her that she would hardly feel anything; he stopped himself from telling her that most of the pain would be experienced by the donors.

Rotten teeth were the price the wealthy of Georgian England paid for their lifestyle. These days everything seemed to have sugar in it – from tea at breakfast to the sweets many sucked before bedtime. All this sugar was ruining the nation's smiles. If this lady ever hoped to find a husband, something would have to be done. She could have had some false teeth made – carved for her from ivory – but these rarely fitted well. No, thought Spence, in coming to him she had made the best decision.

Across Europe tooth transplants had been carried out for many years. The surgical textbooks gave detailed accounts of how to carry out transplantation operations, with some suggesting the use of animal teeth. Ambroise Paré (see Chapter 1) was one of the many eminent surgeons who wrote about the procedure, describing the case of a noblewoman who received a tooth transplant from one of her ladies-in-waiting. By 1780, transplanting teeth from poor donors to wealthy recipients had become commonplace. There were a few voices claiming that it was morally dubious, and others who came to realize that the transplants were rarely successful, but Spence backed neither of these views.

Spence went to examine the women queuing by the back door. Some of them he dismissed straight away, including one whose face was covered in sores and another who looked like she was in need of a tooth transplant herself (not that she would ever be able to afford it). The remaining three he ushered inside so that he could take a closer look. They all appeared to be in reasonable health – no telltale signs of venereal disease or TB. Spence remained unconvinced that disease could be passed on from the donors to recipients. Still, best to be on the safe side. He explained to the three women what was going to happen and how much they would be paid for their contribution. His servant went to fetch the pliers.

Spence seated his first donor down on a couch in a back room and asked her again if she was willing to go through with the extraction. She nervously agreed. Spence took the pliers and gripped an upper left canine. With his knee placed against the couch for leverage, he pulled sharply on the tooth, twisting it until it came away. The woman screamed as blood poured from her mouth and dribbled down her chin. The servant passed her a handkerchief to plug the wound and Spence headed next door to perform the transplant.

He instructed his patient to open her mouth and try not to make a sound. She grasped her friend's hand as Spence held her jaw steady. With the pliers he extracted one of her rotten teeth and with his lancet made a slit in her gums. He gave the new tooth a quick wipe with a cloth to remove the blood and jammed it into the cavity. The patient was sobbing with the pain but did her best to keep her mouth open. Her friend reassured her that she was being brave.

Spence finished off by looping a thread around the new tooth and tying it to the adjacent teeth. The first transplantation completed, Spence went back to his donors for more teeth.

After an hour or so, the donors left with a few shillings and fewer teeth, and the patient nursed a swollen jaw. Nevertheless, she was very pleased with the outcome – a glance in the mirror was all she needed to be convinced that all the pain had been worthwhile. Spence assured his patient that the swelling would soon subside and congratulated himself on another successful operation.

Within a few days the swelling had indeed subsided, although the new teeth felt a little loose. After a fortnight she started to experience sores around her mouth and a rash developed across her body. A physician was sent for, but when a large abscess began to eat away at her nose it was obvious to everyone that she was suffering from syphilis. It could only have come from the teeth; the disease must have been passed on from the donor's blood. Within a few months the whole side of her once beautiful face was horribly disfigured. It wasn't long before the poor young woman was dead. All because she had wanted some nice white teeth.

Spence is said to have infected at least seven of his wealthy patients with syphilis. But it wasn't incidents like this that finally put a stop to tooth transplantation, or the fact that almost all tooth transplants failed through rejection. What brought the practice to an end was the invention of an alternative: ceramic false teeth. However, the idea that living matter could be taken from one person to be transplanted in another was an idea that was far too good to dismiss for long.

Although Spence's forays into transplantation often ended in disaster, his reputation as a dentist attracted the attention of a young surgeon, John Hunter, a man obsessed with understanding what made something alive – the 'living principal'. It is difficult to know how best to describe Hunter. Pioneering surgeon, teacher, naturalist, philosopher – he was to become all these things. His observations and 'scientific' experiments shed new light on biological processes. He advanced the understanding of the human body, both its anatomy and physiology, and devised daring new medical and surgical techniques. His collaboration with Spence not only yielded the first accurate scientific study into teeth, it also gave Hunter far more ambitious ideas.

After witnessing tooth transplants, Hunter started dabbling with other transplantation experiments. He cut the spur from the foot of a cockerel and grafted it on to its head; he took a human tooth and transplanted it on to a cockerel's comb; he even transplanted the testes from a cockerel and attached them to a hen. In a few cases his transplantation operations appeared to be completely successful, but most of them failed. It is thought that the transplants between animals succeeded only because, through inbreeding, the chickens were genetically very similar.

Hunter showed that transplantation was possible – albeit a little hit or miss – and made the first tentative steps towards understanding it. Future generations of dentists, surgeons and scientists, including Charles Darwin and Joseph Lister, would owe Hunter an immense debt of gratitude. People would visit the museum he founded and marvel at his scientific discoveries. But while Hunter would be commemorated as a great pioneer, other surgeons who pushed the limits of science would not be so lucky.

A CAREER THAT STARTS WITH MURDER

Lyon, 25 June 1894

French president Sadi Carnot had rarely received such a rapturous welcome – not only from the mayor and city officials of Lyon (which was only to be expected), but from the exuberant crowds that filled the streets everywhere he went. During the last few days, horses draped with the flag of the republic had led his carriage to banquets held in his honour, and from a balcony he had watched a torchlight procession and a display of fireworks and illuminations. He had marvelled at the wonderful exhibits on view at the Exhibition of Arts, Sciences and Industries. Finally, after all the excitement, he was looking forward to an evening at the theatre, where a gala performance had been arranged.

The president left the Lyon Chamber of Commerce, where he was guest of honour at yet another banquet, a few minutes after nine o'clock. Thousands of spectators cheered as he crossed the short distance to his open carriage waiting outside. Everyone was trying to get close, pressing to catch a glimpse of the French leader. The president didn't mind – it was wonderful to be greeted in such a way.

As he settled into his seat and the carriage started to move off, a young man in a light brown suit and peaked cap was pressing his way through the crowd. He was clutching a newspaper in his hand, but no one took much notice of him; he was just another person jostling for the best view. Suddenly, the man jumped on to the carriage step and flung aside the newspaper to reveal a dagger. The president barely had time to react before the knife was plunged into the left side of his chest and he slumped back against the seat.

President Sadi Carnot was still alive, but unconscious. The assassin had hardly withdrawn the dagger before he was seized by the crowd, their cheers having turned within seconds to screams of horror. The man was punched to the ground amid cries that he should be killed there and then. As the police did their best to protect him from the fury of what had now become an angry mob, the carriage containing the dying president was rushing towards the hospital.

As the assassin was bundled off to the police station, protected by police and mounted guards, the president was laid on a bed. His condition was worsening. A crimson stain on his shirt was spreading as blood seeped from the wound and dripped on to the sheets. Lyon's finest surgeon was summoned. At the police station the assassin gave his name as Cesare Giovanni Santo, a twenty-two-year-old Italian anarchist with a poor grasp of French and, as one reporter put it, 'a very small moustache'.

The doctors who had now gathered around the president concluded that the dagger had missed his heart, but one of his major blood vessels 'in the region of the liver' had been punctured. The blood pouring from his wound told only part of the story: most of the blood was being lost inside the president's body.

They did their best to stem the flow with towels and bandages, and at 11.30 the surgeons issued an optimistic statement in which they stated that the condition of the president was 'alarming but not hopeless'. The bulletin went on to suggest that the haemorrhage had ceased. The surgeons were wrong. The president was bleeding to death and there was nothing anyone could do. At 12.45 the president of France was declared dead.

French justice was swift. A little over a month after the assassination the president's murderer was tried, convicted and executed. The Lyon surgeons concluded that Carnot had died from blood loss caused by a wound to the portal vein – the major blood vessel from the intestines to the liver. Even if they had attempted to operate on him, they would have had little chance of success. No one had managed to mend a broken blood vessel before; the only option open to surgeons was to tie off blood vessels completely. This was fine in a limb, although cutting off the blood supply inevitably ended in amputation. When it came to a major internal vein or artery, the procedure was out of the question.

The whole affair was deeply shocking, particularly for a young doctor at the hospital, Alexis Carrel. In his autobiography Carrel wrote how the president's life 'left him with his blood, in the midst of the holiday crowd. I can still hear it flowing drop by drop fifty years later.'* Carrel was appalled that the surgeons had been unable to save the president. The death set him on a path that would lay the foundations for modern transplant surgery. It would lead Carrel to a Nobel prize, a partnership with the world's most famous aviator, and into collaboration with the Nazis.

Alexis Carrel was an odd-looking young man. He had the countenance of someone who thought about things a lot, whether it was the death of a president, the latest surgical advances or the future of

* *It is very unlikely that Carrel was at the bedside of the dying president, but in later years he seems to have convinced himself – and everyone else – that he was. There is also some debate over exactly where the president died. Some historians claim it was at the hospital, whereas reports in newspapers of the time say it was in the city* préfecture.

humanity. Carrel was intense, self-absorbed and somewhat distant. It was as if he were observing the world from a higher intellectual plain. Physically, he was also quite unusual. Anyone meeting him for the first time found it difficult to identify what it was about his appearance that was so unsettling, until they peered into his tiny eyes. Behind his pince-nez, which he wore for his chronic short-sightedness, you could see that one of his eyes was brown and the other one blue.

Throughout his life Carrel collected enemies, usually power ful ones. And where better to start his collection than with his superiors at the hospital in Lyon – those surgical butchers who considered themselves such experts, but who had so impotently failed to save the life of the president. They said saving him had been impossible; Carrel thought they were wrong, said as much and set out to prove it.

First he needed to develop his sewing skills, but not the brutal, clumsy sewing he was used to seeing during operations in the hospital (he sometimes wondered if surgeons took pride in the grotesque size of the scars they left behind). If Carrel wanted to sew together delicate blood vessels, he needed to learn how to do minute, delicate, precise stitching. So he headed to Lyon's silk district and acquired the services of the city's finest embroiderer, Madame Leroudier.

Carrel was a driven and conscientious pupil and, with the smallest of needles and finest silk thread, he worked tirelessly to perfect his technique. Just trying to thread the microscopic eye of an embroidery needle takes considerable patience and determination but Carrel worked night after night, much to the derision of his macho medical colleagues. Within months, he had not only mastered the stitches, but was almost as good as Madame Leroudier herself. It was said that Carrel was so proficient that he could place five hundred tiny stitches in a single piece of cigarette paper. Now all he had to do was apply his beautiful embroidery technique to some veins.

Blood vessels are circular, slippery and easily damaged. Cut one and it resembles a damp, floppy drinking straw. Clamp a vein or artery with forceps and it is left crushed and bruised; sew a vein back together again and it will almost certainly leak or clot – either way, it will be all but useless. Carrel had to overcome all these problems if he was to sew blood vessels together successfully, so he headed back to the laboratory with his tiny curved embroidery needles and fine silk thread and set to work.

The first thing he worked out was how to stop the flow of blood without damaging the vessel. Using bands of cloth, he would gently squeeze vessels shut and successfully hold back the blood. If he rolled back the edges of a cut vessels so that they resembled cuffs, he could sew the cuffs together without the usual leaks or damage that led to clots. But it was Carrel's final discovery that was his most masterful. He called it the 'triangulation method' of suture.

First, he joined the ends of the blood vessel together by placing three stitches equal distances apart around its edge. For each stitch he left a short piece of silk thread attached. The blood vessel was thus joined at three points. Now here's the clever bit: when he pulled tight on all the threads at the same time he created a straight line between each one. He had turned the circular vessel into a triangle. He could then sew along the straight lines between the threads. It was a remarkably simple idea, but extremely effective. He had overcome the problem of trying to sew around a circular vessel by doing away with the circle. Once he had sewn along the first line, he moved on to the second and the third. When he released the threads, he was left with a neat sutured join in the blood vessel. It was such a simple technique that even the surgeons he held in contempt would be able to manage it.

It was no great surprise to anyone when Carrel failed to receive promotion at the hospital in Lyon, and it was probably best for all concerned when he left for the United States to pursue his research. In Chicago he teamed up with Charles Guthrie, a similarly obsessed medical researcher, and together they improved Carrel's technique with ever finer needles and thread. They stitched severed veins and arteries, and joined the two together. Usually the veins and arteries belonged to a dog, sometimes a cat, or occasionally a guinea pig. Some animals survived the procedures, some didn't. Carrel was unsentimental about any creatures that might suffer in the name of medical progress.* Armed with his new surgical techniques, he had a much greater purpose in mind: transplantation.

INSIDE THE LABORATORY OF ALEXIS CARREL

Rockefeller Institute for Medical Research, New York, June 1938

Dr Alexis Carrel was now considered one of the world's most eminent and famous scientists. Surgeons, politicians and celebrities came to seek his advice or expert opinion. This month he was even

* *The animals would have been anaesthetized, and there is no evidence that they were mistreated. Whether the experiments themselves amounted to mistreatment is a matter for debate.*

on the front page of *Time* magazine. The reporter had been lucky to secure an interview – Carrel was usually hidden away in his laboratory, guarded by his protective secretary, as devoted to his work as ever. Since winning a Nobel prize in 1912 for his work on 'Suture of Blood Vessels and Transplantation of Organs', Carrel had conducted thousands of transplantation experiments. He had transplanted limbs from dogs, kidneys from cats and testicles from rabbits. He had taken lungs from guinea pigs, heads from dogs, ovaries from cats and thyroids from kittens (the attrition rate of kittens was particularly high). He had swapped the leg of a black dog with the leg of a white dog and replaced the head of one dog with the head of another.* He had grafted kidneys, livers and lungs; he had transplanted organs, glands and legs. He had swapped skin, rearranged veins and added hearts. No animal, it seemed, was safe from Carrel's increasingly bizarre research.

Carrel's laboratories were every bit as sinister as the experiments. Built on the top floor of the Rockefeller Institute, they were reached from an anteroom by a narrow spiral staircase. Here Carrel's fanatical team of researchers worked in sterile windowless labs. They were lit by roof lights and electric bulbs in the plainest of shades. Everything else was in varying tones of black – from the matt black floors to the bare grey walls. Even the cloths on the operating tables were black. There was no colour, no

* *Carrel wasn't alone. In 1954 a Russian transplant surgeon, V.P. Demikhov, went even further, transplanting the head of one dog on to the back of another to create a monstrous two-headed creature. This disturbing experiment was brought to an end after the two heads started fighting each other.*

dust and no reflections. Dracula couldn't have conceived a more suitable lair.

The outfit the scientists wore might have been designed by a fetishist. The researchers were known as the Black Gang. They worked in black gowns and black trousers. Their heads were covered in black linen balaclavas. The headgear was square in shape – like a welder's helmet – with only a narrow slit for the eyes. On their hands the scientists wore thick, black rubber gloves. These shadowy, featureless men would be the last thing most of their experimental animals ever saw.

There was a good reason for all the black. Carrel had designed the labs to minimize reflection and glare from the lights – vital, he believed, to help the researchers see what they were doing, particularly when they were working on such tiny bodies (such as those of kittens, rabbits and guinea pigs). The all-enveloping outfits were designed to minimize infection. Joseph Lister would have been proud of the lengths to which Carrel had gone to keep the place aseptically clean (see Chapter 1), although Lister might have preferred a cheerier colour. The odd thing was that the only one who wasn't dressed entirely in black was Carrel himself. He had taken to wearing a peculiar white hat that resembled a bandage stretched over his bald head.

In the late 1930s most of Carrel's laboratory was devoted to his 'perfusion' experiments. He had gone beyond simply transplanting organs to removing them altogether. His aim was to keep organs – eventually human ones – alive and functioning in a totally artificial environment. For the last three years he had been working with his close friend, Charles A. Lindbergh, the first man to fly across the

Atlantic. Although it might seem an odd partnership, the celebrity aviator and eccentric French scientist had much in common. Lindbergh shared many of Carrel's political views, they were similarly driven, and both were ambitious to advance medical science.

In 1938 Carrel and Lindbergh were celebrating the publication of a book they had written together, *The Culture of Organs.* In it they outlined the 'cultivation' of organs using the Lindbergh pump. The pump was designed to bathe living tissue in nutrients to keep it alive, and looked exactly like the sort of thing a scientist working in a sinister black laboratory would devise. The contraption consisted of a series of pumps, bottles, gauges, valves and odd-shaped glass flasks all connected with lengths of rubber tubing.

The pump, the culmination of years of effort, was really an early type of heart-lung machine – it kept organs alive, nourishing them and providing them with oxygen. At around the same time, surgeon John Gibbon (see Chapter 2) was also developing a heart-lung machine, only his aim was to keep entire organisms alive – eventually humans – while life-saving surgery was carried out. By the late 1930s Gibbon could sustain the life of a cat. In comparison, Carrel's motives were far more scientifically detached. He was able to sustain the life of a disembodied cat's heart. Improving treatments or saving people's lives wasn't enough for him. He had far loftier ambitions.

For a start, the Lindbergh pump was to be used to study organs outside the body. It allowed Carrel to examine the nutritional requirements of a particular organ, or study the chemical processes taking place. He could assess the production of insulin in the pancreas, urine excretion in the kidneys or the life cycle of cells. With the techniques he developed, Carrel took the creation of 'cultures' of living tissue to a whole new level. On one bench in his black laboratories he grew cells taken from the heart of a chicken embryo. So far he had managed to keep successive generations of these cells alive for twenty-six years.

This was all very well, but it was only incidental to his life's work on transplants. Carrel planned to remove damaged organs from a patient's body – a diseased kidney, for instance – place the organ in the pump apparatus and treat it in this artificial medium until it healed. Once the organ had been cured, he would replace it in the patient using the techniques he had first developed in Lyon. It didn't even need to be a kidney – it could be a leg, an arm or even a brain. If Carrel's laboratory was nightmarish now, imagine what it would be like lined with bottles of bubbling glass jars filled with dismembered limbs, diseased hearts and cancerous lungs.

But even this wasn't enough for Carrel. In what he called his 'new era' of surgery he foresaw a time when human organs could be grown in the lab and used to manufacture drugs such as insulin. He never made it clear who might provide these organs, but the ethics of his work always came second to scientific progress. As the reporter from *Time* magazine suggested, Dr Carrel was 'looking for the fountain of abundant, replaceable age'. But probably not for everyone.

Carrel's collaboration with Lindbergh underlay a much deeper moral purpose. The two men were out to change the human race. Three years before, in 1935, Carrel had published his philosophical treatise *Man, the Unknown.* It was widely read and drew acclaim from scientists, statesmen and intellectuals around the world. In it Carrel outlines his views on everything from future scientific progress to the role of women in society. He was convinced that man was in a state of physical, mental and moral decline and needed to be 'remade'.

'For the first time in the history of humanity, a crumbling civilization is capable of discerning the causes of its decay,' Carrel wrote. 'For the first time, it has at its disposal the gigantic strength of science.' The recent political and economic turmoil had demonstrated the failings of democracy; he wanted to see a new social order. Countries should be run by a ruling elite, their standing in society determined by biological worth. Man had the power to transform himself, to control his genetic destiny. Only the strong should be genetically perpetuated.

The irony that a short, balding, myopic Frenchman (with different-coloured eyes) should be calling for the creation of a master race was lost on Carrel. But his call for the introduction of eugenics was well received. The idea that only the genetically 'superior' should be allowed to breed (or encouraged to breed) was something that many in power had been thinking for some years. At various times everyone from Winston Churchill to George Bernard Shaw and H.G. Wells had flirted with the philosophy of eugenics. In six American states laws had been in place for decades to allow the forced sterilization of the insane and 'mentally deficient', and in Germany Carrel's scientific standing added credibility to the philosophy of the Nazi government. In the German edition of his book, he even went so far as to endorse Nazi policies.

Carrel probably had little knowledge of what was really going on in Nazi Germany, despite having travelled and lectured there in 1936. By the mid-1930s eugenics was at the heart of German government policy. Hitler was oppressing opposition groups, people of Jewish faith, ethnic minorities, homosexuals and gypsies (among others). The government had set up concentration camps, as well as secret extermination centres where the mentally and physically disabled were being killed. At the same time women of Aryan descent were being encouraged to have as many children as possible.

However, while Carrel advocated preventing the criminal or insane from breeding, he shied away from destroying 'sickly or defective children as we do the weaklings in a litter of puppies'. Instead he felt the only way to 'obviate the disastrous predominance of the weak is to develop the strong'. Among his suggestions was the proposal to remove the sons of rich men from their families so that they could 'manifest their hereditary strength'. And although he believed childless women were 'not so well balanced', unlike many misogynists he was a firm believer that women should be highly educated, 'not in order to become doctors, lawyers or professors, but to rear their offspring to be valuable human beings'. As for his treatment of criminals, those who couldn't be conditioned with a whip should be 'humanely and economically disposed of in small euthanasic institutions supplied with proper gases'. You can see why the Nazis were so taken with his views.

Not everyone was impressed. There were rumours that 'occult practices' were taking place in Carrel's black laboratories. Even that Lindbergh was planning to have his heart removed so that it could be replaced by a mechanical device of Carrel's creation. This wasn't so far from the truth: the two men *were* operating in a scientific hinterland. Lindbergh was fascinated by questions of life and death, and had contemplated immortality. Carrel was a believer in clairvoyance and telepathy, and an advocate of the power of prayer (he even wrote an article about it for the *Reader's Digest*).

On 28 June 1938 Carrel was sixty-five years old and the strict rules of the Rockefeller Institute meant that he was forced to retire. For someone so passionate about his work, this was incredibly frustrating, but perhaps now was the time to realize his life's ambition and set up his own institute devoted to the study of man. An institute that would build on his theories outlined in *Man, the Unknown*. An establishment that would set mankind on a triumphant path to the future.

Carrel's opportunity came in February 1941 (some ten months before the United States entered the Second World War) when he joined a relief mission to take food and medical aid to occupied France. Quite what Carrel's motives were for joining such an endeavour is unclear, but it wasn't long before he'd offered his services to the Vichy government. He saw the downfall of France as evidence for his theories on society. France had been 'crushed because of our corruption, vanity and weakness'. Now he had the opportunity to help in 'remaking' the country.

On 17 November, despite the fact that the French economy was crippled by the German occupation and most ordinary people were on a starvation diet, Carrel was awarded a generous budget to set up his new institution. His foundation would study measures to 'safeguard, improve and develop the French population'. Offices were commandeered at the Rockefeller Foundation in Paris and, seemingly oblivious to the suffering taking place around him, the self-absorbed little Frenchman resumed his experiments.

After Paris was liberated by the Allied forces in August 1944, no one was sure whether or not Carrel should be arrested as a collaborator. Although his work was backed by the hated Vichy authorities and endorsed by the Nazi leadership, Carrel had merely been getting on with his research. During the war he had even spoken out against the inadequate rationing imposed by the Germans, and counted members of the Resistance among his friends. Some efforts were made to detain him, following press accusations that he was a pro-Nazi racist (which was, strictly speaking, true). The American ambassador was even asked to intervene on his behalf. In the end Carrel died before anyone could decide what to do with him. Sympathetic biographers have claimed that he died of sorrow, devastated that people thought so ill of him. Within a few years his public image had gone from scientific hero to Nazi villain.*

Alexis Carrel could have been a great medical hero; instead his name has been all but erased from the history books. His enormously popular text *Man, the Unknown* was removed from most libraries, his perfusion experiments abandoned, his laboratories shut down. Scientists and surgeons were embarrassed to have been associated with him. The press that once sang his praises no longer mentioned him. Everyone conveniently forgot that they had once thought eugenics was a good idea. It was left to a few loyal

* *Charles Lindbergh almost suffered a similar fate. In 1941 he gave a speech in Iowa, during which he called for appeasement with Germany. His anti-British and anti-Semitic views were widely vilified, and many people, including Lindbergh's own mother-in-law, distanced themselves from him. After Japan attacked Pearl Harbor in December that year, Lindbergh returned to aviation, flying more than fifty combat missions in the Pacific and instructing many young pilots. However, it was several years before his reputation was sufficiently rehabilitated for him to once again be considered an American hero.*

colleagues (including Charles Lindbergh) to try to put Carrel's side of the story. As a result, the biographies are deeply divided. Some are damning in their condemnation, others are obsequious in their praise. But if you try to look beyond the man to his many achievements, they are quite remarkable.

Carrel was the first doctor to work out how to sew blood vessels together. This discovery alone helped save countless lives. When arteries or veins were damaged, his efforts meant that they could be successfully repaired. His technique made transplants possible and opened up a whole new area of surgery. His pioneering experiments with tissue culture gave scientists a much greater understanding of organ and cell function. They also allowed the investigation of conditions such as diabetes. His idea that limbs and organs might be grown or repaired in the laboratory was decades ahead of its time. Advances in stem cell research might one day make this possible. As for his views on eugenics, it is hardly fair to single out Carrel as a villain. In the first half of the twentieth century his views on the future of humanity were shared by many other influential people.

However, for all his achievements and technical advances, Carrel kept coming up against a major problem. Transplants of an organ between different parts of the same animal were invariably successful, but almost every one of his hundreds of transplants between different animals ultimately ended in failure. The operations had gone well, the organs would function for a while and then, within days (or occasionally weeks), they would fail. Carrel concluded he was coming up against a biological force that he was powerless to counteract. Despite fifty years of experiments, he failed to overcome a major obstacle to successful transplantation: rejection – the body's reaction to foreign tissue. In his battle with the body's immune system, he was defeated time and time again.

The immune system is in a constant state of war and, as Carrel discovered, a transplanted organ provides an easy target. Everything alien to the body comes under attack, from transplanted tissue to bacteria, fungi and viruses. The human body employs a whole range of different cells and techniques to repel invaders, and the immune system is continuously evolving to adapt to new threats. Transplant an organ and the body will rapidly set to work to try and eliminate the foreign tissue. The immune system even has its own distribution network – the lymphatic system – and a series of lymph nodes where the various immune system cells congregate.

The white blood cells form the core of the body's immune response. Although they are called white blood cells, they are actually transparent. Every single millimetre of blood has some ten thousand white blood cells, all poised to take on invaders. The first line of defence is made up of neutrophils, which can swallow up and kill bacteria. These are backed up by the even more fearsome lymphocytes, which come in two main forms: B and T.

B and T cells are manufactured in the bone marrow, found in the long, flat bones of the body, such as the pelvis. B cells produce fragments of protein called antibodies, which bind to the surface of foreign invaders. These antibodies either disable the invader or mark them out for destruction. T cells come in two different varieties – helper T cells and killer T cells. The helpers work with the B cells to produce antibodies and also assist the development of the killer T cells. It is these killer T cells that are the really nasty ones.

They target anything identified by the B cells as alien, ambush the invaders and destroy them.

Doctors know all this now, but fifty years ago the body's immune response was still shrouded in mystery. For transplant surgery to be successful, the formidable barriers of the body's own defences would have to be studied, analysed and overcome. In the meantime, some surgeons were prepared to carry on regardless.

LIFE AFTER DEATH IN FRANCE

Paris, 12 January 1951

Seven years after Alexis Carrel's death another Frenchman was about to die. He was scheduled to be executed by guillotine within the walls of the Santé prison in Paris. Not that many people cared. Ever since the abolition of public executions in 1939, interest in the death sentence had waned considerably. This was merely another routine execution of a criminal that society could probably do without.

It was a bitterly cold morning. The executioner's breath mingled with the icy morning mist and choking smog of the city. The guillotine stood in the courtyard, blocked in by the towering brick walls of the prison. What a place to die.

The killing machine itself was once considered the height of technology – a machine to end life efficiently and humanely. These days it was beginning to show its age. The plank where prisoners rested their bodies for the last time was worn, the high wooden gantry was discoloured from age, and even the bucket for catching the disembodied head was looking battered. The only part that still appeared as good as new was the glistening steel blade, sharpened the previous day.

The executioner examined the ropes on the machine and checked the straps on the plank. He positioned the bucket of sawdust where he judged the severed head would fall. With public executions it used to be embarrassing to see a head bounce from the basket and roll towards the crowd. It wasn't dignified. The least an executioner could be was professional. He pulled on the rope and hauled the heavy blade to the top of the gantry, then fastened it before releasing a lever to let it go. Satisfied that it was working properly, he hauled it up again. Now went to see how the prisoner was getting on.

The condemned man had been given the last rites. The irony of the situation sometimes made the executioner smile. It crossed his mind that if the man – this criminal – had really believed in God, he wouldn't have committed the crime in the first place. Still, who was he to judge? He was only doing his job and the priest was only doing his. It was best not to think about it too much, particularly in this profession.

A guard tied the prisoner's hands together behind his back and led him from the cell. The man shivered slightly as he was taken into the courtyard. Some prisoners struggled, but this one seemed as calm as could be. There was no point resisting the inevitable; it achieved nothing and only made the whole thing more unpleasant for everyone.

The prisoner was pushed forward on to the plank and his head placed in the semicircular groove of the 'lunette'. The executioner fastened the straps around him and instructed him to lie still. Two men standing over by the wall turned their heads away, slightly embarrassed witnesses. They had other thoughts on their minds. The executioner checked once again that the bucket was in place and the prisoner was positioned correctly. He told his colleague to stand clear and moved towards the lever. Everyone was silent.

There was a click. The blade dropped so rapidly that its movement was barely perceptible. The head dropped into the bucket with a soft thud – eleven pounds of brain, bone, muscle and skin gently oozing into the sawdust. It steamed in the cold air. Where the neck had been severed a great arc of blood spurted out. The fountain gradually subsided to a gentle trickle, congealing on the frosty ground.

The two men who had been waiting and trying not to look (although in truth it was almost impossible not to) took this as their cue. The headless corpse was carried inside to a table and its clothes cut off. The men put on their masks and gloves and, working quickly, sliced open the warm body. They weren't too careful with their incisions – the prison would clear up the mess afterwards. However, they needed to be sure not to damage the kidneys they were trying to remove. Within minutes, they had what they wanted. Dousing the organs in fluid designed to keep them alive (similar to the fluids Carrel used in his experiments), they wrapped them in towels and headed for the hospital. These two men – surgeons Charles Dubost and Marcel Servelle – planned to make transplant history.

In the operating room their first patient was being prepared for surgery. The forty-four-year-old woman was lying anaesthetized on the operating table, cloths draped across her, nurses ready with trays of instruments. The bright light and pristine surfaces were in marked contrast to the shabby conditions at the prison. When the surgeons arrived with the dead man's kidneys the organs were doused in more fluid to wash them and prevent them deteriorating. While Dubost and Servelle scrubbed for surgery, one of the kidneys was brought to the operating theatre for transplant.

The surgeons implanted the kidney into a cavity in the woman's pelvis – connecting it into the pelvic blood vessels. The ureter – the tube leaving the kidney, which normally carried urine to the bladder – was passed through a hole in the skin. On the same day, they carried out an identical operation on a twenty-two-year-old woman using the other kidney from the executed prisoner.

At first both operations appeared to have been successful. Within two hours of receiving the transplant, the older woman began to excrete urine from her new kidney. Over the next few days the volume of urine increased. The second patient seemed to be recovering equally well. Perhaps Carrel was wrong; perhaps they could overcome the body's defences? The surgical team was cautiously optimistic, even allowing themselves a low-key celebration.

The forty-four-year-old woman died seventeen days after the operation. The younger woman died suddenly after nineteen days. In both cases the transplanted kidneys had been destroyed by the immune system. The Paris surgeons went on to perform a total of eight transplant operations. They used kidneys from living donors; they washed the kidneys before transplant; they used the best available medication and provided round-the-clock intensive care. Despite all their efforts, every one of the eight patients died (although one lasted more than a month). In each case the new organs seemed to be incompatible. The biological force Carrel had warned about continued to defeat them. But there was every reason to keep trying. The patients the Parisian surgeons operated on were all in the final stages of kidney, or renal, failure. Without functioning kidneys they would certainly die slow, unpleasant deaths.

Kidneys act as filters to the blood. They remove waste products from the body to produce urine. They also help to maintain the right balance of fluids and regulate blood pressure, hormones, minerals and red blood cells (among other functions). The first symptoms of kidney failure include lethargy, nausea and swelling of the ankles as a result of a build-up of fluid. Without treatment, symptoms progress through nausea and breathlessness to confusion, seizures, blindness and eventually coma. It's not called 'end stage' renal failure for nothing.

Unfortunately, there were few effective treatments for acute renal failure in the 1950s. The only alternative to transplants was dialysis, but few hospitals offered it at that time. Dialysis used an artificial membrane to filter waste products from the blood. The process had been invented in the 1920s and developed by a Dutch physician, Willem Kolff, during the Second World War. Kolff's machine consisted of a large tank, cellophane tubing (made from sausage casing) and a rotating drum that resembled a paddle from an old steamboat. This artificial kidney was hooked up to the patient with rubber tubing and the motor switched on.

Kolff's first patient, a twenty-nine-year-old woman, showed dramatic improvements in her condition following dialysis. When she had been admitted to the hospital her eyesight was failing, her heart was enlarged and her breathing was laboured. After dialysis her vision and breathing returned to normal and, reported Kolff, 'her mind was perfectly clear'.

The problem was that every time the doctor needed to use the artificial kidney, he had to cut into major arteries and veins where the blood pressure was strong. He could insert glass tubes into a patient's arms, upper legs, even neck, but each time he did so the blood vessels were irreparably damaged (doctors use the word 'exhausted'). Each site on the body could be used only once, so patients could be attached to the artificial kidney only so many times. Eventually, Kolff ran out of suitable blood vessels. His first patient underwent twelve treatments, but despite his best efforts, she eventually died. Although he had proved that dialysis worked, he had merely prolonged the life of his patients for a few weeks or months, not cured them.

After the war, a few hospitals in Europe and the United States adopted Kolff's technology or built new types of dialysis machine. However, they all ran up against the same problem. Dialysis was difficult, cumbersome and often dangerous. It was a last resort to keep people alive.* Surgeons needed an alternative, and kidney transplants still looked like the best bet. But with only a limited understanding of the immune system, how could they overcome the problems of rejection?

Surgeons tried everything they could think of. One surgeon had the idea of transplanting a kidney wrapped in a plastic bag. The

* *The problems of dialysis were not solved until the 1960s, when a device made of new types of artificial tubing (a combination of Teflon and plastic) was developed. This 'shunt' was permanently connected to the patient's blood vessels so that they could be easily and repeatedly attached to the dialysis machine.*

theory was that the bag would create a barrier against the immune system. The patient survived for six months, but the relative success of the operation was thought to have little to do with the plastic bag. Surgeons suspected that the reason the kidney had lasted so long was that the patient was reasonably well matched to the donor. This seemed to be the key – if the donor and recipient could be matched for blood, tissue type and immunity, the transplant would probably be successful.

In Boston, Massachusetts, the surgeons at Peter Bent Brigham Hospital had been working on the problems of kidney transplants for many years and were becoming increasingly disheartened. Would they ever manage a successful transplant? Finally, in 1954, they hit on some extraordinary good luck.

THE IMPORTANCE OF SHARING

Boston, Massachusetts, October 1954

Richard Herrick was in a terrible state. Since the twenty-three-year-old had been admitted on 26 October he had caused nothing but problems for the staff. He had knocked over equipment and pulled out his catheter. He had cursed doctors and accused them of sexual assault. He had even bitten one poor nurse on the hand while she was trying to change his bedclothes. In the end he was moved to a side room to keep him from disturbing the other patients.

None of this was Richard's fault. He was in the advanced stages of kidney failure, and his psychotic behaviour was its most pronounced symptom. He was only dimly aware of his surroundings, he could no longer recognize people, had little idea where he was and only a tenuous grasp on *who* he was.

Richard had been referred to the Peter Bent Brigham Hospital as a last resort. If anyone could save his life, it was the surgeons here – the most experienced transplant surgeons in the world. That said, they had yet to perform a single kidney transplant operation with any long-term success. However, Richard hadn't been admitted just because of kidney failure – there was no shortage of equally deserving cases – but because he had an unusual biological quirk. He had a twin brother, Ronald, who was willing to donate one of his own kidneys.

The surgeons knew from previous experiments that transplants could be carried out between identical twins. They had tried transplanting small skin grafts with some success. Identical twins seemed to share the same immune system. Now they had the perfect opportunity to try it out with a kidney. This was a case of the right patient in the right place at the right time. Transplant surgeon Dr Joseph Murray called it 'happenstance favouring a prepared mind'. However, before making the decision to go ahead with the operation the doctors wanted to be doubly sure that the brothers were indeed completely identical twins.

Richard was given dialysis to stabilize his condition, and the surgeons performed every test they could think of. They drew samples of blood to check the blood groups matched. They did. They rang up the brothers' family doctor to see if they had shared the same placenta in the womb. They had. The surgeons examined their eyes to see if they were exactly the same colour. They were. Murray even took the brothers down to the local police station to have the detectives check whether their fingerprints were identical.

In all, the surgeons carried out some seventeen tests and the brothers passed every one, but they would have to wait a month for the results of the most crucial test. Murray had transplanted a small piece of skin from Ronald to Richard. If the graft were successful, the surgeons would be in a position to make the final decision as to whether to go ahead with a kidney transplant.

The pressure on the surgeons to operate was building. The press had got wind of the transplant operation. When the brothers had been fingerprinted, crime reporters hanging around the police station had started asking questions. Soon the news was all over the newspapers. But the surgeons could deal with the media; the bigger problem came from Richard himself. After more than a month in hospital his condition was, once again, deteriorating. Despite the dialysis, Richard's heart was starting to fail. His death would be only a matter of time.

Ronald visited him every day at the hospital. The family knew that Richard wasn't going to make it. The surgeons were certain that his death was imminent. If the operation went ahead, there was every chance that Richard could die on the operating table. Even Ronald started to have second thoughts. He was young and healthy – what were his own chances in life if he gave up a kidney? Ronald loved his brother more than anyone else in the world (both their parents were dead), but what if they both died during the operation? Having a kidney removed was in itself a major operation. Would the surgeon who operated on *him* be competent and experienced? After a lot of soul-searching, Ronald came to the conclusion that he would go ahead and donate his kidney. Then, despite knowing he would definitely die without the operation, Richard tried to persuade his brother to pull out. He even wrote a note telling him to get out of the hospital and go home. But Ronald had made up his mind.

Even the surgeons were beginning to wonder if this was the right thing to do. They had been assured that removing a kidney from a healthy adult had no adverse long-term effect on health or life expectancy. Nevertheless, they consulted psychiatrists, lawyers and even local clergy. Was it morally and ethically right to remove a perfectly healthy kidney from a living donor? Richard was becoming sicker by the day and time was running out. With the skin graft showing no signs of rejection and with Ronald's full consent the surgeons eventually reached a decision.

JOINED AT THE HIP

Peter Bent Brigham Hospital, 23 December 1954

The two operating theatres are next door to each other. Ronald and Richard Herrick both lie unconscious, shrouded in linen sheets, their bodies illuminated by the bright operating-theatre lights. Each of the twins is surrounded by a team of masked surgeons, nurses and anaesthetists. Every conceivable instrument that might be required is laid out ready. Drips are set up for blood and plasma transfusions; there are swabs, needles, knives and tweezers. The surgeons have practised on cadavers. Joseph Murray has worked through the operation a thousand times in his head. At 8.15 a.m. he is ready to start.

The surgeons removing the kidney from Ronald feel the strain as much as Murray. This is the only compatible kidney on the planet. If they mess it up, Richard will die and they could put Ronald's life at risk. Each team works slowly and carefully, Dr J. Hartwell Harrison on Ronald, Dr Murray on Richard. By 9.50, Harrison has exposed the blood vessels leading to Ronald's kidney. He is ready to sever the blood supply and remove the organ. In the operating theatre next door Murray has prepared the site in Richard's pelvis where the kidney will be reconnected. Everyone pauses. Murray takes a deep breath and gives the instruction to remove Ronald's kidney.

At exactly 9.53 a.m. the surgeons wrap the donated kidney in a cold wet towel and carry it through to Murray's operating theatre. Murray knows he has to reattach the severed kidney as quickly as possible to re-establish the blood supply. The fist-sized organ is sitting in a stainless-steel bowl. Who knows how long it will last?

The clock is running.

Murray has already clamped off the iliac artery in Richard's pelvis at the very top of his right leg. Now he begins to sew. As Carrel had discovered, joining together blood vessels is a slow and precise procedure, but half an hour later the surgeon has successfully connected the artery of Ronald's donor kidney to the artery in Richard's leg.

Murray works methodically and precisely. Everyone is anxious. Is he taking too long? He tries not to look up at the clock. It is 10.40 a.m.

Now he needs to connect the vein from the kidney to the vein in Richard's leg. It is slow work but he can't get distracted by the clock. After thirty-five minutes the veins are joined.

Murray makes a final check to see if everything is OK. Ronald's donated kidney has been out of his body – and without a blood supply – for a total of one hour and twenty-two minutes. Will it still work?

Everyone goes quiet; they can hardly breathe with the tension. The surgeons gently loosen the clamps around the blood vessels. The blood begins to flow. The transplanted kidney becomes engorged. It turns pink, pulsing with blood.

There is a collective sigh of relief; Murray even allows himself a smile. Within minutes, urine starts spurting from a catheter on to the floor. They mop it up and connect the ureter to Richard's bladder. The transplanted kidney is working perfectly.

The next day Richard is feeling better than he has for months. His eyes are bright, he is alert and hungry. Richard and Ronald are discharged from hospital in February. They are both fit and healthy. X-rays confirm that Richard's new kidney is functioning well. As the newspapers put it, this surgery was truly a 'medical miracle'.

Richard went on to marry his nurse and father two children (not identical twins). He lived for another eight years, eventually dying of a recurrence of kidney disease. The surgeons had proved that with identical twins the immune system could be beaten. Over the next few years they tried the procedure on several more sets of twins with equal success.

The surgical techniques developed by the Boston team continue to be used to this day in the tens of thousands of kidney transplant operations that take place every year. But twins represented only a tiny proportion of the people who needed kidney transplants. Understandably, Murray wanted to treat all his patients.

He wanted to be able to offer a kidney transplant to anyone in need. The only way to do this was to take on the immune system, and he believed he had just the thing.

THE NUCLEAR OPTION

United States, 1957

Welcome to the atomic age, where nuclear energy makes everything possible. Why not vacation in Las Vegas – the 'Atomic City' – to see the awesome power of the atom for yourself? While you're there, you could get yourself an atomic hairdo and head off to a 'Dawn Bomb Party' in the desert to witness the latest nuclear test. You could take an atomic box lunch before heading back to the city to sip an atomic cocktail while watching the Miss Atomic Bomb contest. You might even get to see the lucky winner posing in her dazzling white mushroom cloud outfit.

Nuclear tests were the biggest thing that had ever happened to tourism in Nevada, and the crowds flocked from all parts of the USA to see the flash, feel the heat and witness the cloud. But it wasn't just in the desert that the atomic age was capturing the imagination. Right across America there was talk of nuclear-powered rockets and cars. Every home would soon have its own nuclear reactor; housewives would preserve and cook food with the wonders of atomic rays. The US military was spending some $70 million a year on a nuclear-powered aeroplane (although it still had to overcome a few issues with safety). The dream of cheap, clean, nuclear energy was being realized. Nuclear was the future and the future was now!

Medicine was no stranger to the wonders of the atom. X-rays had revolutionized diagnosis and allowed surgeons to see moving images of the inside of the body (see Chapter 2). Radiation was also being used to treat cancer, helping to save many thousands of lives a year. Other doctors were studying the biological effects of radiation – vital to refine treatments, ensure people's safety and, of course, plan for the aftermath of a Third World War.

Ever since the first atomic bombs were dropped on Japan in 1945, scientists had been building up a better understanding of the effects of radiation on the human body. Doctors had been able to examine victims of radiation sickness as their symptoms progressed from vomiting, diarrhoea and fatigue to the full-blown and invariably terminal signs – hair loss, uncontrolled bleeding and heart failure. They found that some parts of the body were affected more than others, and when scientists started to look at individual cells (often during post-mortems) they concluded that some cells were more sensitive to radiation than others. The most vulnerable cells turned out to be those that line the intestine (hence the vomiting and diarrhoea) and also the cells of the immune system – the white blood cells. Enough radiation and the immune system could be completely wiped out. This discovery got the transplant surgeons thinking. Could radiation be used to overcome the body's defences and break down the barrier to successful organ transplants?

A few experiments were tried on animals with varying degrees of success, but despite some misgivings, the surgeons at Peter Bent Brigham Hospital in Boston decided to go ahead and treat their transplant patients with radiation. Joseph Murray planned to use X-rays to suppress his patients' immune system before conducting a transplant. This would, in theory, avoid rejection and allow the transplant to 'take'. Their patients had nothing to lose – they were going to die anyway – so any new idea was worth a try.

The first patient was thirty-one-year-old Gladys Loman. A mother of two, she had been born with only one kidney. When it became infected it was accidentally removed in an emergency operation. The surgeon responsible thought he was removing a diseased appendix. This left Loman with no kidney at all and only weeks to live. She was referred to Joseph Murray, who gave her dialysis to keep her alive. This, he warned her, could be used only a few times. After that she would either die or he could attempt his experimental new procedure.

Gladys Loman lay on a mattress beneath the X-ray machine. She was curled up in a foetal position to expose her immune system to the radiation. The X-rays would destroy the white cells in her spleen, lymph nodes and bone marrow. Radiation would wipe out her body's defences and leave her completely vulnerable to the slightest infection. The surgeons turned on the machine and left the room for their own safety. Gladys lay on the mattress for three hours trying not to move. Above her the X-ray tubes bombarded her with massive amounts of radiation.

Following the procedure he had adopted with Richard and Ronald Herrick, Murray transplanted a healthy kidney into Gladys. The kidney had been taken from a stranger and would normally have been used by researchers at the hospital who were studying polio. Gladys's new kidney was completely alien to her body. The question was, now that her immune system had been destroyed, would she accept the new organ?

To avoid the risk of infection, Gladys was housed in a completely clean room – actually a converted operating theatre. When anyone came to see her they had to scrub their hands and wear operating gowns, hats, masks and gloves. She couldn't leave – she was trapped in this sterile hospital prison.

At first the new kidney failed to work, but eventually, after two weeks, it started to produce urine. It looked like the operation had been a success, so the surgeons gave Gladys a bone marrow transplant to try to give her immune system a boost. But her body had had enough. Thirty days after the transplant operation she was dead.

Gladys had endured dialysis, major surgery and massive doses of radiation. She had coped with pain and discomfort, and spent a month isolated from the world in an operating theatre. All that for a few extra days of life. You have to wonder whether it was worth it.

Staff at the hospital were becoming more and more despondent, and one surgeon quit altogether. Despite his own misgivings, Murray still believed the immune system could be overcome, and pressed on with the total irradiation procedure for eleven more patients.

By the third patient, twenty-six-year-old John Riteris, the surgical team had refined the procedure. Instead of administering the radiation in a single large dose, they used the X-ray treatments in shorter bursts. They studied cases of people involved in nuclear accidents and looked again at data from animal experiments. With Riteris, it helped that the kidney donor was the patient's brother; they were twins, but not identical twins. Their differences were revealed when a skin graft between them was rejected. Nevertheless, the surgeons reckoned that they still might have a better chance of success.

Riteris's new kidney worked almost immediately. Although his white blood cell count was alarmingly low, he managed to remain free of infection. Better still, it looked like the organ wasn't being rejected. Over the next few months he was given further doses of radiation, as well as anti-inflammatory drugs. Eventually, he left hospital with a working kidney and went on to lead a normal, healthy life. At last the surgeons had broken another barrier – they had shown it was possible to transplant organs between non-identical brothers.

But any triumph was short-lived. All the remaining transplant patients who received total body irradiation treatments at the hospital died. Radiation suppressed the immune system, but in doing so it laid the patients wide open to infection. It was the infection that killed them. The atomic dream was over. Surgeons needed to look for something else.

THE MAGIC MUSHROOM

Cambridge, England, 1976

The odds on surviving a transplant operation were improving every year, but they still weren't great. By 1965 around four out of five transplant operations were successful if the donor and recipient were related. If they weren't related, the odds fell to around one in two. In the UK in the early 1960s one of the world's most experienced transplant surgeons had conducted a series of fourteen kidney transplant operations. Only one patient survived.

There had, however, been a number of innovations during the 1960s that made transplants more likely to succeed. Surgeons were now able to match the immune systems of the donor and recipient more closely. This process, known as tissue typing, greatly improved the odds on the transplanted organ being accepted. And, with total body irradiation abandoned, scientists had developed new drugs to help combat the immune system's defences. Still, going into hospital for a transplant operation could be a grim experience, especially for children. A nine-year-old girl admitted to the Royal Infirmary in Edinburgh in 1967 later described how she was kept in isolation to avoid the risk of infection. For the five weeks following her kidney transplant operation, the only people she came into contact with were masked nurses and doctors, who had to scrub and shower before entering and leaving her room. The girl's parents were barred from entering, and could only communicate with her through a window.*

As for the new drugs, they came with a substantial health warning and a list of side effects that ran to a small dictionary. They might suppress the immune system, but the A–Z of nasty things these drugs could also do to the body ranged from the inconvenient to the fatal: from alopecia to tremor, anaemia to ulcers, cancer (through heart disease and nausea) to osteoporosis. A common side effect of, for instance, the steroids being prescribed, was facial swelling – a syndrome known as 'moon face'.

Even with the drugs and the tough procedures to keep patients isolated from infection, too many transplant patients were dying. But this didn't stop surgeons trying new and daring operations. By 1970 they had moved on from transplanting kidneys to livers and

* *Fortunately, the discomfort was worth it. The girl's kidney was still working more than thirty years later.*

the pancreas. They had even transplanted a human heart (see Chapter 2). But organ transplants were increasingly perceived as the last desperate measure of an increasingly desperate branch of surgery. Many hospitals refused to carry out transplants – it hardly helped their mortality figures. Murray later described the period at the end of the 1960s as 'transplantation's darkest hour'.

Then surgeons had a stroke of luck. Jean Borel, a young researcher at the Swiss drug company Sandoz, was given the task of examining a bag of Norwegian soil. The soil had been gathered during an expedition to a bleak highland plateau, and it was Borel's job to see if he could find anything useful in it. After careful analysis he was rewarded with the discovery of a new type of fungus from which he extracted a chemical. They called it cyclosporine A. This was no new penicillin – cyclosporine A was useless at killing bacteria – but it did appear to have a dramatic effect on the immune system. Borel found that cyclosporine suppressed the function of the T cells (specifically the helper T cells), preventing the immune system from attacking alien tissue.

In 1976 Borel attended an English surgical conference to report his findings. When the transplant surgeon Roy Calne heard about the remarkable new substance Borel had discovered, he couldn't wait to get his hands on it. Calne had been one of the pioneers of transplant surgery, and among the first to use drugs to suppress the immune system. He had teamed up with Murray in the 1960s and had been instrumental in improving the success of organ transplants. Now Calne wanted the opportunity to try cyclosporine. Could this drug finally provide the breakthrough surgeons so desperately needed?

Calne persuaded Sandoz to send him a sample of cyclosporine so that he could try some experiments for himself. But when the sample turned up there was a major snag: it was in its purest form – as a white powder – and the researchers in Calne's Cambridge laboratory couldn't get it to dissolve. Neither water nor any of the other common solvents they had lying around the lab worked. This meant that if cyclosporine was made into a pill, it wouldn't be absorbed in the gut. As a drug, it was all but useless. Sprinkling some white powder on the transplanted tissue to see what happened wasn't really an option. In the end the future of transplantation surgery was saved by a protective mother. Alkis Kostakis's mother to be precise.

Kostakis was a visiting research fellow from Greece, but his mother was worried. She was particularly worried about English food, and with good reason. In 1976 English cuisine was, as a rule, lurid, processed and bland. Even the blandest of English foods, the potato, now came in freeze-dried granules; green vegetables were boiled to slime; and Angel Delight – a mousse-like substance with an indeterminate flavour – was considered a sophisticated dessert. Orange juice (in bottles) was a once-a-week treat and bread (white, sliced) had all the nutrients baked out of it as a matter of course. No wonder Mrs Kostakis was worried.

She sent her son a bottle of finest Greek extra virgin olive oil. But before Alkis could drip it on to his salad (iceberg lettuce was probably the best he could hope for), he took the oil into the lab. More in hope than expectation, he decided to try mixing it with the cyclosporine. He had nothing to lose, so he carefully measured out the oil and ladled it over the precious fungus powder. The drug dissolved. He tried out his combination of olive oil and cyclosporine on a series of animal patients. The results were spectacular. They were so spectacular that Calne didn't believe him, so he sent Kostakis back to repeat them. But he got the same results again – cyclosporine mixed with olive oil worked wonders. Soon Calne could start trials in human patients; he would transform the world of transplant surgery. All he had to do was lay his hands on more cyclosporine.

But Sandoz, the company that had discovered cyclosporine, was not convinced. The way things had been going with transplant surgery in recent years, they saw no future in cyclosporine. As far as they were concerned, it would only lose them money. Calne flew out to see them. He talked to their financial people, he argued, he cajoled, he badgered. He told them this was the best thing he had seen in all his years of transplantation. Finally, Sandoz gave in and agreed to conduct a limited drugs trial. They were still reluctant. It would, they warned, almost certainly lose them money.

Surgeons began testing the drug on transplant patients in 1978. With the introduction of cyclosporine, survival rates rocketed. One year after their liver transplant operations some 70 per cent of patients were alive, and almost 80 per cent after kidney transplants. Cyclosporine wasn't without its own side effects, and the risk of infection was still a major problem, but it looked like the immune system had finally been overcome.

In 1990 Joseph Murray was awarded a Nobel prize for his work on organ and cell transplantation (he shared the award with E. Donnall Thomas, who had developed drugs to minimize rejection). Roy Calne was knighted for his transplant research in 1986 and is one of the few surgeons to be elected a fellow of the Royal Society.

In the bloodstained history of surgery, transplants stand out as an area where even the best surgeons have been defeated time and time again. From Spence's disastrous tooth transplants to Carrel's sinister laboratories, experiments with decapitated French criminals and total body irradiation, transplantation surgery is littered with ill-conceived ideas, gruesome experiments and procedures bordering on the unethical. It took until the mid-1980s for transplant surgery to become a safe, routine surgical treatment. After more than two hundred years the battle with the body's own defences had been won. Now anything was possible. Hearts, livers, lungs, kidneys; surgeons could even transplant a dead man's hand.

CHAPTER 4
FIXING FACES

THE ITALIAN NOSE JOB

Bologna, Italy, 1597

Bologna was fast becoming the syphilis capital of Europe. This wasn't something anyone advertised or put on the signs. It wasn't good for business – particularly if your business was prostitution – but the ravages of the disease were clear for all to see. Syphilis was debilitating, disfiguring and, in most people's opinion, downright disgusting.

An unwelcome import into Italy from South America, syphilis is caused by tiny coiled bacteria. The disease is spread through contact and, as it needs moisture to survive, the contact is often of a sexual nature. Without treatment, syphilis spreads rapidly through the body. It starts with swelling around the site of the infection, but within weeks the victims develop rashes, fevers and headaches. They will suffer painful lesions in the mouth, throat and anus. As the disease progresses, the body becomes covered with ulcers and tumours, and clumps of hair fall from the head.

Worse is to come.

While the patient becomes increasingly disfigured on the outside, the bacteria are conducting a hidden campaign of destruction inside the body. They attack bones and muscles, covering them with rubbery tumours that affect posture and movement. If the victim has somehow managed to conceal the effects of the disease up to this point, syphilis then launches a final nasty surprise. As these tumours spread, they begin to erode the bones of the nose. When the nose collapses, the victims are left deformed, their face distorted, their appearance repulsive.

Doctors had all sorts of treatment on offer for syphilis. These invariably involved bloodletting or expensive concoctions of herbs and unlikely bits of animals. Nothing was effective. Within a matter of weeks, the victim went from upright citizen to social pariah, with a caved-in face to match. Sufferers were shunned as moral degenerates. They would do anything to have their faces restored. Here was the perfect market opportunity for any enterprising surgeon.

Gaspare Tagliacozzi was undoubtedly one of Italy's greatest surgeons, renowned as a brilliant practitioner. He had risen rapidly through the ranks of the University of Bologna Medical School – Italy's foremost medical university. By the age of thirty-five he had been honoured with civil office, and had even been granted the privilege of conducting public demonstrations of anatomy. As his reputation spread, Tagliacozzi's rich, famous and, importantly, influential clients came to include the very finest of Italian nobility.

One of his earliest patients was the distinguished Count Paolo Emilio Boschetti of Modena. The count had suffered a broken arm that had healed badly. He came to Tagliacozzi seeking treatment for the stiffness.

The surgeon examined the limb and diagnosed that there was a problem with the 'materials within'. So that movement could be restored, the tendons and ligaments needed to be softened. Tagliacozzi had been schooled well and knew just what was needed. He prescribed that the arm should be held in the warm entrails of a sheep for an hour a day. Afterwards the arm was to be placed in a hot bath of herbs for half an hour. Finally, it should be washed with warm wine before being dried. It was important that the patient had not previously eaten anything, so perhaps the count might consider undertaking his treatment before breakfast? Although whether he would feel like having breakfast after dousing his arm in bits of dead sheep is debatable.

Tagliacozzi's treatments were in the finest traditions of the self-appointed father of surgery, Claudius Galen (see Chapter 1). Despite their dubious efficacy, they were well received by patients, and Tagliacozzi soon had a thriving business, in addition to his salary from the university. But while private clients made him wealthy, it was his anatomical demonstrations that drew the crowds and helped make his name as a surgeon. He had a reputation as a fine teacher and commanded great loyalty among his students.

Anatomical dissections were undertaken only by senior members of the faculty. They were such rare events that the lecture theatres were usually packed and the doors guarded by four of the 'most quiet and serious students'.* Their job was to make sure that only students, doctors and perhaps, if there was room, 'those persons of good qualities' entered the theatre. There had been a few problems in the past with troublemakers from the lower orders getting in (there had also been a few cases where enterprising students had demanded payment from gawpers wanting to be admitted). The authorities were keen to stress that these were events for learning, not common entertainment.

The cadaver – a criminal allocated by the city authorities – was laid out on a slab at the centre of the room. The dissection was performed in constant reference to Galen's texts, and took place over a period of several days. Incisions were made and organs removed in strict order. The whole event was as much ceremony as lecture, with enough religious overtones to keep the powerful Church authorities happy. Tagliacozzi became so proficient at dissections that he was soon appointed professor of anatomy. It is therefore surprising that such a disciple of Galen and pillar of the surgical establishment should also turn out to be a great surgical innovator.

Tagliacozzi was fascinated with the idea that a damaged face might be restored. He started to develop a new branch of medicine: what he called the surgery of 'defective parts'. Although syphilis was one of the most prevalent causes of facial disfigurement, it was not the only way people could lose their noses. It was not unusual for them to be severed on the battlefield or in a duel. Unfortunately, even if their noses had been hacked off in an honourable way,

* *This was according to the official decree that detailed the strict rules governing public dissections.*

syphilis had become so prevalent that people confused victims of the sword with the sinful victims of syphilis. More and more people were coming to Tagliacozzi desperate for a new face, but any attempt at reconstructive surgery was fraught with difficulties.

To be fair to Tagliacozzi, *any* surgery in the sixteenth century was fraught with difficulties. First, any operation had to be conducted without anaesthetic. Patients generally only agreed to the pain if the alternative was death, so surgery might be considered for a life-threatening condition such as a gangrenous leg wound. But could surgery be justified if it was only to restore a person's appearance?

The second problem was infection – the slightest cut in the skin could become infected and ultimately kill the patient. The only incisions surgeons made on a regular basis were for bloodletting; otherwise they preferred to stick with external treatments involving herbs, spices and possibly entrails. Overriding both these considerations was the problem of technique. If a surgeon were going to rebuild a nose, where was the skin going to come from? Attempts had been made to take skin grafts from donors (slaves or servants usually), but these had always been unsuccessful. The skin had to come from somewhere else on the patient's body. Tagliacozzi chose to take it from the arm.

However, it turned out you could not simply cut into the arm, remove a slice of skin and stitch it to the face; the patch would wither and die. There was also a good chance that the wound left on the arm would become infected. To remain viable, the skin had to remain connected to a blood supply. Tagliacozzi's solution was not without considerable pain, inconvenience and embarrassment for his patients (not to mention cost), but it was simple.

PROFESSOR TAGLIACOZZI INVENTS A CURIOUS OUTFIT

The patient had not left his house for many months. The shame of being seen in public would have been too much to bear. His face was shocking to look at – where he had once had a nose there were only two scarred hollow sockets. Even his wife made every excuse not to see him, although given that he lost his nose through syphilis, this was hardly surprising. Whereas only a few months ago the man had been out every day, he now lived for the most part in his bedchamber, visited by only a few trusted servants. It was a grim existence, and one that he hoped Professor Tagliacozzi would be able to rectify. Otherwise, he believed he would probably take his own life.

Tagliacozzi's knife is razor sharp, his movements rapid and precise. He slices the blade into the patient's flesh on the underside of the upper arm, making a cut about as long as a nose. He removes the knife and makes a further cut parallel to the first. He then makes a cut between the top of the two lines. The knife is so sharp and the incisions so quick that the patient feels hardly any pain. The cuts redden as blood seeps out. It drizzles down the man's arm and drips into a bowl on the floor. Tagliacozzi mops the wound with a handkerchief and moves on to the next stage of the operation. Sorry, sir, but this part is going to hurt.

The surgeon slips his knife through one of the cuts and passes it horizontally underneath the skin. The patient screams in agony as Tagliacozzi runs the knife backwards and forwards between the two parallel incisions. He slices through nerves, blood vessels and fat, gradually lifting the skin as he goes, pulling it away from the underlying tissue. Now the pain is becoming unbearable. The man is desperate for this terrible torture to end. Tagliacozzi's assistant struggles to keep the patient's arm still. It takes only a few minutes for the surgeon to finish, but for the man it feels like an eternity.

When all the cutting, slicing and scraping is finally over, the patient is left with a rectangular flap of skin on his arm and a gaping wound. Tagliacozzi carefully lifts the flap with his fingers and dresses the raw tissue underneath with strips of bandage that soon become sodden with blood. The raised skin, known as a pedicle, remains connected to its blood supply at the lower end of the rectangle, although the exposed edges are already healing. Now the surgeon needs to graft the skin to his patient's face.

When the patient raises his arm in front of his face, the pedicle rests across the empty sockets of the nose. Being connected to the patient's arm, the pedicle is supplied with blood and, with the help of a few stitches, will grow into the man's face. When the thousands of tiny capillaries and veins in the face have made their connections, the pedicle can be severed and the finishing touches put to the new nose. The problem is that it takes at least two weeks for the new blood supply to be established. In the meantime, the patient has to hold his arm across his face.

Try holding your arm up in front of your face so that your upper arm rests on your nose. Now try holding it there for two minutes. Hurts, doesn't it? Imagine how it feels to hold it for two hours. Or even two days. To get around this impossibly uncomfortable situation, Tagliacozzi designed a novel item of headgear. It consisted of a leather corset and helmet supporting a series of belts and straps. The straps held the patient's arm in place so that the hand rested on the back of their head. Their wrist was attached to the helmet to restrain movement, and straps around their head prevented the arm swinging from side to side and accidentally ripping the pedicle.

Tagliacozzi had this peculiar bondage outfit tailor-made for each patient. Once on, it had to remain on for two weeks – the patient's hand strapped across the top of their head, their elbow jutting out in front of their face, their movement and vision restricted. It was cumbersome and looked ridiculous, but people were prepared to try anything to restore their features.

The jacket and headdress were only part of Tagliacozzi's elaborate treatment plan. As the pedicle gradually started to grow into the stump of the nose, the surgeon insisted that his patients follow a strict diet. They were allowed meat – but it should be roasted, not boiled – and he advised that they avoid fish. At least there weren't any entrails involved. With the straps securely tightened on the corset, the patient was left groaning on his (or her) bed.

A fortnight later the surgeon returns to see how the patient is getting on. By now the top of the pedicle has grown into his nose. The tissue is still healthy and Tagliacozzi can sever the connection between the upper arm and the face. A quick slice with the knife and, much to his relief, the patient can remove the leather jacket and lower his arm.

After two weeks he, like most of Tagliacozzi's patients, finds his muscles so cramped that he can barely move. The stench when he takes off the leather jacket is somewhat overpowering. As for his appearance, if anything it has got worse. Where he had once had half a nose, he now has a flap of skin dangling in the middle of his face. In true Renaissance fashion, Tagliacozzi needed to become an artist.

Using splints, bandages and the occasional stitch, the surgeon starts to rebuild the nose. Over the next few weeks, he slowly sculpts his patient's new face. Three months after the first incision, the skin has grown together, the splints have done their job and the bandages can be removed. Carefully pulling out the final splint, Tagliacozzi holds up a mirror. His patient's new nose is revealed in its full glory. Slightly scarred and somewhat different in colour from the rest of his face, it is still a considerable surgical achievement. He can once again go out in public. Tagliacozzi was truly a miracle worker.

The surgeon published the first-ever book on reconstructive surgery in 1597. Within it he outlined his methods and included detailed diagrams to illustrate the various stages of nasal and other types of facial reconstruction. The techniques he devised would remain familiar to surgeons well into the twentieth century.

Unfortunately, after Tagliacozzi died in 1599 his reputation collapsed. The Italian Church had been growing suspicious of his activities. Now that he was in no position to defend himself, the Church summoned its investigation team: the tribunal of the Inquisition.

Tagliacozzi was accused of magical practices. He had modified the human face and in doing so had been interfering with the will of God. In the end the Church allowed his soul to rest in peace, although stories persist that Tagliacozzi's body was removed from its tomb and his bones dumped on unconsecrated ground.

At the time Tagliacozzi's method was a major advance on anything that had gone before, although his techniques built on more than two thousand years of surgical practice. The first recorded case of plastic surgery took place in India around 1500 BC. The Hindu epic poem *Ramayana* tells the story of Surpanakha, a beautiful temptress (some say a demon with magical powers). With her bewitching personality, Surpanakha attempts to seduce a young prince who is promised to another. She is sentenced to a brutal punishment for her actions and her nose is cut off. However, this is far from the end of the story. Rather than live with the disfigurement, she goes for reconstructive surgery.

An Indian medical text dated to around 600 BC gives an idea of the sort of treatment Surpanakha would have received. First, the doctor would have cut a nose-shaped flap in her forehead – narrow at the bottom, above the nasal cavity, and wide at the top. The incision would have been around a quarter of an inch deep, down to the periosteum, the thin fibrous membrane covering the skull.* The doctor would then have peeled the skin away from her forehead, making sure not to tear the narrow part at the bottom. This strip of skin, rich in blood vessels, would become the pedicle and keep the skin flap alive. Twist the pedicle around and bend it down and there you have a new nose. You also have excruciating pain and an appalling (nose-shaped) scar on the forehead. It was a crude technique, but better than having no nose at all.

Surprisingly, despite Tagliacozzi's advances, the cruder Indian technique was still being practised by surgeons well into the nineteenth century. Seemingly reluctant to try any surgery that took

* *Periosteum membrane covers all bones, but the forehead is one of the few places on the body where the skin is right against the bone. Periosteum contains tough fibres of collagen and nerves, as well as blood vessels to supply the bone cells.*

more than a few seconds, Robert Liston (see Chapter 1) dismissed the Italian method as too tricky. The Indian operation, on the other hand, was 'less difficult in execution, not so liable to failure, and more easily undergone by the patient'.

In his book *Elements of Surgery*, Liston describes in detail his own variation on what he termed the 'rhinoplastic operations'. Liston suggests making a wax mould of the nose and then flattening it out so that it becomes a template for the skin flap. However, he confesses that this can be a difficult process and it is often more convenient to use a piece of cardboard (you can guess which method he used).

The card was held firmly by an assistant as the surgeon traced around it with a pen, 'or at once with a knife carried deeply through the integuments'. It is hard to imagine Liston bothering with a pen first. With the template removed, Liston describes pulling the skin away from the forehead using his finger and thumb. If it becomes difficult, he suggests the use of a hook. Finally, the flap is twisted around and placed over the area of the nose, the wound in the forehead is dressed and a couple of straws are stuck up the nostrils so the patient can breathe. Understandably, many people opted for false tie-on noses rather than endure the horrors of Victorian surgery.

However, it was another Liston innovation that revolutionized plastic surgery: anaesthetic. Before pain relief, surgery was the last resort of a desperate patient – whether it was to remove a diseased limb or fix a disfigured face. Now, though, a whole glorious new world of surgery was about to open up. People were no longer coming to surgeons to fix their faces: they wanted to *improve* their faces. Nose jobs, smaller breasts, facelifts or bigger lips – there was nothing surgeons wouldn't try. And with infection defeated by antiseptic techniques, operations were becoming much safer.

A new era of cosmetic surgery had arrived, and surgeons (some more qualified than others) were, once again, in the exciting business of experimenting on their patients. Bizarrely, rather than perfecting operations to move flaps of skin around, they developed operations that involved inserting a whole range of novel substances beneath the skin. It seemed there were few products of the Industrial Revolution that weren't brought to the operating table. Surgeons attempted rebuilding noses with ivory, they experimented with metal, celluloid and gutta percha (a substance derived from tree sap); they tried oil and coal extracts; even bits of animal cartilage. One surgeon brought a live duck into the operating theatre, slit its throat and attempted to repair his patient's nose with the bird's breastbone. They notched up their failures to experience until, finally, they hit upon the perfect new substance.

GLADYS DEACON: A CAUTIONARY SURGICAL TALE

Paris, 1903

Twenty-two-year-old Gladys Deacon lay in bed contemplating her own beauty. She was, undeniably, exceptionally beautiful. She was also extraordinarily vain.

Intelligent, charming and wonderful company, Gladys was all these things and more. Why, hadn't a young gentleman told her that this very evening? He was handsome certainly, but a mere plaything to Gladys, who had set her sights on marrying into royalty (or landed gentry at the very least). Still, it was nice to be admired; although few people could come as close to admiring Gladys as much as Gladys herself.

Raised in Boston, Massachusetts, she moved in all the right circles. She then burst on to the European social scene, mixing with aristocrats and artists, princes and politicians. A friend talked of how Gladys traversed Europe 'like a meteor in a flash of dazzling beauty'. The press adored her, men courted her, other ladies envied her. She was becoming famous for being famous – a true Edwardian celebrity. But as she lay on her bed thinking about herself, she started to have doubts. Could she possibly become more beautiful?

Like many Edwardians, Gladys was fascinated with the classics and the concept of classical beauty. She toured the galleries and museums of London, Paris and Rome, examining statues and studying paintings. She admired the profiles of Hellenic faces; their strong foreheads and straight noses. She even took to recording the distances between the eyes and noses of statues to see how they measured up. But when she compared herself with them, her observations brought her to an alarming conclusion: she wasn't perfect after all. Her nose dipped between her forehead and the tip, creating a slight hollow. She wanted a straight classical nose, and she knew just what to do to get one.

Gladys went to see a professor at the Institut de Beauté in Paris. He examined her and advised that she try the latest advance in cosmetic surgery: paraffin wax. Unlike previous innovations, the wonder of this new treatment (invented only a few years before) was that there was no actual surgery involved. All the surgeon had to do was inject a measure of hot paraffin wax under the skin: as it hardened, he could mould and shape it to create the perfect profile. Paraffin wax had been injected into faces, breasts, buttocks and even the occasional penis. It really was a remarkable invention. It sounded almost too good to be true.

The surgeon wears thick, black rubber gloves as he prepares the paraffin. The solid white block of wax is gradually turning to a slushy liquid as he heats it on a small oil burner. The large glass syringe, with its formidable wide steel needle, is lying alongside in a basin of hot water. The surgeon has learnt from experience that unless the syringe is also hot (hotter by several degrees than the wax), the paraffin will solidify before it can be successfully injected. Reaching over, he checks the temperature of the paraffin with a thermometer. It is a careful balance: too cold and he can't inject it, too hot and it will burn the patient. The ideal temperature is around 30°C, but it is difficult to get it just right. He has heard of cases where the skin has simply sloughed away from the patient's face, presumably due to excessive heat. Still, there are risks with all types of surgery.

Gladys lies back on the couch. She has tied back her hair to expose her beautiful, smooth, (near) perfect face. The surgeon's assistant dabs the bridge of the young lady's nose with some dilute carbolic acid to clean her skin. The surgeon sits on a stool beside her, the pan of hot melted paraffin wax and warm syringe at the ready. She gasps as he makes a small nick in her nose with a scalpel. He places the tip of the needle in the hot wax and draws the asbestos piston (rubber would have melted) of the syringe upwards to fill it. Even with gloves on, the surgeon can feel the heat as he places his fingers through the loops at the top of the syringe and prepares for the injection.

Gladys is proud of how brave she is being. She has been warned that it will be painful, but pain is surely a small price to pay for perfection. As the surgeon sticks in the broad needle and depresses the plunger, Gladys feels as though molten metal is being injected into her head. The paraffin wax squirts out through the needle and beneath the skin of her nose. The surgeon keeps pressing until the syringe is almost empty, then he flings it aside and begins to mould Gladys's new face.

He has between fifteen and thirty seconds to get it right. The fingers of his bulbous gloves push, knead and press. He glances down at a picture Gladys has provided so that he can check his work. He runs his fingers along her nose, smoothing any bumps, moulding the paraffin like putty beneath her skin. The paraffin wax is hardening rapidly and time is running out. The surgeon presses as hard as he can to stop the wax clumping. A few seconds later and it has set; but he is finished. Gladys Deacon has a new nose.

The surgeon explains to Gladys there may be some swelling at first, but this will soon disappear. In just a few days, he tells her, she will have a classical nose to be proud of. He applies a compress of lint dipped in icy water to numb the pain and sends her home.

The swelling was indeed quite bad to start with. Only it didn't get any better. Instead it got worse, the bridge of her nose bursting into an angry open sore. Doctors were summoned to examine her, but when she was questioned Gladys denied having had any surgery. Instead she blamed her inflamed features on an accident, telling people she must have knocked it. But the nose got worse: the wax began to wander; lumps appeared beneath her skin. Her beauty was slowly being destroyed from within. Far from achieving the classic looks of a Greek statue, her quest for perfection was turning her into a freakish waxworks dummy.

It was little consolation to Gladys that she wasn't alone. Despite the ringing endorsement of many eminent surgeons, including England's Stephen Paget, who recommended the use of paraffin wax in the *British Medical Journal*,* others had begun to notice that these injections often led to unwanted side effects. In fact, the list of side effects was alarmingly extensive. The condition was even given a name: paraffinoma, although some doctors simply called it wax cancer.

Symptoms ranged from the odd lump to wide abscesses where skin withered and died. Paraffinoma caused infection and destroyed muscle. If the paraffin got into the bloodstream it led to blood clots in the lungs and was held responsible for blindness, strokes and heart attacks. The price of perfection was quite possibly death.

In his 1911 book on plastic surgery, American surgeon Frederick Kolle highlighted the dangers of paraffin wax injections. He also warned doctors against the 'unreasonable' demands made by patients who were 'bent upon having the alabaster cheek ideal of the poets, the nose of a Venus, the chin of an Apollo'. He

* *Stephen Paget was considered one of Britain's finest surgeons. In a gushing article in the September 1902 edition of the* British Medical Journal, *he described how paraffin wax was simple to use and produced excellent results. In his own practice, he said, the outcome was 'absolutely satisfactory'; he even gave the name of the company from which the paraffin wax could be purchased. To be fair, the wax did sometimes produce excellent results but, given that no one had carried out any proper trials, it is impossible to know what proportion of injections was successful and what proportion ended in disaster.*

referred to these people as 'beauty cranks' – those seeking perhaps 'very desirable marriages'. Surprisingly, it seems he had never met Gladys Deacon.

By the 1920s the wax injection had really taken its toll on poor vain Gladys. She wore a hat low over her face to disguise the worst ravages of paraffinoma, but female rivals recorded bitchily how the wax had given her face the appearance of a gorgon. Others remarked that she looked heavy jawed, her hair too yellow, her lips too red. She no longer looked like a lady (the implication being that she looked more like a whore). A princess who had once been jealous of Gladys noted with ill-disguised satisfaction how the wax had run down her face to create blotchy patches in her neck.

However, while society mocked her for her medical mistake (behind her back, of course), Gladys continued her climb up the steps of the social ladder. In 1921 she finally made it into the British aristocracy by marrying the 9th Duke of Marlborough and taking up residence at Blenheim Palace in Oxfordshire. But despite being a duchess, she was becoming more and more depressed.

By the 1940s, her marriage having failed, she was to be found living in a ramshackle farmhouse. She slept on a broken mattress surrounded by the squalor of cats, rotting food, papers and books. Gladys was becoming increasingly frail, isolated and paranoid, and it wasn't long before four men in white coats came to literally drag her away.

Gladys Deacon died in her sleep in a Northampton psychiatric hospital in 1977. The funeral was poorly attended. Most people had forgotten Gladys Deacon, Duchess of Marlborough. People said that in her later years she would sit by the fire, letting the heat of the flames soften the paraffin beneath her skin so that she could move it around her face. Gladys never did get the perfect nose.

THE FACES OF WAR

Queen's Hospital, Sidcup, Kent, 1917

It was difficult to look at Lieutenant William Spreckley without experiencing a feeling of utter revulsion. Even the man himself sometimes wished he had been killed when the bullet hit him. His existence, he felt, was almost a living death. He had been passed from the trenches at Ypres to casualty station to hospital before finally ending up in Sidcup, but he didn't hold out much hope for his chances. He would be disfigured for the rest of his life, shunned by society – perhaps even by his own family.

William had a sad, haunted look in his eyes. Although lucky to be alive, he was feeling sorry for himself. Bullets sliced through whatever material they met – whether it was wood, metal or human flesh. Most of his comrades had been cut down: some were killed instantly, others wounded fatally, the rest permanently disabled. William could remember a bright flash of light but, strangely, experienced little pain. He was stretchered away to the crowded tents of the field hospital, where he expected to be left to die. Instead, over the next few weeks he started to recover. He knew his face was damaged, but the nurses and doctors refused him a mirror. The surgeons stitched him up and nurses changed his dressings. By the time William arrived at Sidcup his wounds had healed well. He was fit and healthy. Everything was fine, except for his face.

Instead of a nose he had an ugly, gaping hole. The skin had grown inwards, and what remained of the interior – red tissue and bone – could be seen through the black hollow. The left side of his face was distorted around the hole; a series of lateral scars had healed to draw down the skin beneath his eye, revealing the lower part of his eyeball. But this was nothing compared to the missing nose.

Queen's in Sidcup was the first hospital in the world dedicated to plastic surgery, and the surroundings couldn't be more different from what William had experienced in the trenches. Built in the grounds of a stately home, the hospital was encircled by gardens and tall trees, and even boasted a beautifully manicured croquet lawn. The single-storey wards, treatment rooms and operating theatres were arranged in a horseshoe shape around a central admissions block. Each ward was designed to hold twenty-six beds and included a veranda so that patients could lie outside in the fresh air to help their convalescence (fresh air was considered vital for recovery).

Queen's was the brainchild of surgeon Harold Gillies. He had entered the war as a junior Red Cross doctor in 1914, and had been horrified by the injuries he saw. But Gillies was even more shocked to discover how little British surgeons were able to do to piece soldiers back together. Their techniques were primitive and wholly inadequate. No one had anticipated the terrible carnage – the faces that were blown apart, the missing noses or jaws, the melted flesh and jagged scars. All the surgeons could do was draw the edges of the wounds together, wait for the scars to heal and post their patients back to the trenches to fight another day.

Gillies decided to dedicate his life to plastic surgery, and taught himself everything there was to know about facial reconstruction. Over the next three years (while continuing to work in hospitals in France and England) he studied obsessively, wading through textbooks and research papers. He even enrolled in an art school so that he could learn how to draw detailed diagrams of his surgery. Eventually, he managed to convince the army medical authorities that they needed a dedicated hospital to treat facial deformities. When Queen's Hospital opened in the summer of 1917, Gillies – now Britain's foremost plastic surgeon – was appointed to run it. He was ready to put his vast knowledge of plastic surgery to the test.

William Spreckley was one of the first patients to be admitted to the new hospital. When Gillies examined the young soldier he decided he could do better than simply give Spreckley a new nose. He wanted to improve on the crude efforts of previous generations of surgeons and give Spreckley a nose that really looked like a nose, not some crude flap of skin twisted down from the forehead or grown from the upper arm. He made careful measurements of Spreckley's face and set about planning a series of intricate operations.

Because Spreckley's nose was missing completely, Gillies planned to re-create both the skin and the cartilage supporting it. Rather than repeat the disastrous experiments of his Victorian predecessors and use animal cartilage or synthetic alternatives, Gillies chose to take the cartilage from elsewhere on his patient's body. After drawing up a complicated set of diagrams and technical notes – he believed in the importance of preparation – he was ready to operate.

In the whitewashed, airy operating theatre, with its powerful electric lighting and enormous picture windows, Lieutenant Spreckley is put to sleep.* Gillies, dressed in his sterilized surgical gown, his hands washed in alcohol and covered with fresh rubber gloves, is ready to make his first incision. He cuts into William's chest. The first part of the operation is ingenious and involves removing a small, rectangular piece of cartilage from the soldier's ribcage. Gillies intends to shape this into the support for the nose.

Cutting the cartilage carefully, he bends it along the middle and then cuts away part of the central section to leave a narrow stem. He is left with a shape that resembles an arrow. It has a wide piece at one end, a narrow shaft and a bow-shaped tip. The wide part will form the bridge of the nose; the lower arrowhead will support the nostrils. Once Gillies is confident that the cartilage is the right shape, he slices open a flap of skin on William's forehead and transplants the cartilage under the surface.

When Spreckley recovered from the operation, he looked even more deformed than when he had first been admitted to the hospital. Instead of a flat forehead – which had been undamaged by the bullet – he now had a pronounced, arrow-shaped bump under his skin. The arrow was pointing diagonally upwards from the centre of

* *A major advance in anaesthetics was made at Queen's Hospital. In 1919 Ivan Magill developed endotracheal intubation – the technique of passing a rubber tube through the patient's nose or mouth to allow the gas to flow directly into the trachea. This was not only a more precise means of delivering an anaesthetic, it also overcame a problem that had plagued reconstructive surgery. During the many hours surgeons spent leaning close to their patients' faces, and therefore the gas intake pipe, they often ended up breathing in the anaesthetic. It was not unknown for surgeons to fall asleep during operations.*

his lower forehead towards the line of his hair. Gillies was, in effect, growing Spreckley a new nose in the middle of his forehead. Several weeks later, once Spreckley's forehead was fully healed, Gillies moved on to the next operation.

Cutting carefully, to leave the cartilage intact, Gillies slices a flap of skin from Spreckley's forehead. Making certain not to damage the pedicle, he twists the skin around to form a new nose. The cartilage keeps the structure from collapsing, although the resulting protrusion is hardly attractive. Beneath the angry, triangular-shaped scar on Spreckley's forehead, the new nose balloons out across the soldier's face. He has gone from having no nose to having a swollen, comic representation of a nose. It is horrible. Other patients joke that Gillies has transplanted a trunk in the middle of the poor man's face. Even the surgeon himself remarks in his case notes that 'the new, bloated columella stuck ahead like an anteater's snout and all my colleagues roared with laughter'. But the surgeon is far from finished.

The operations continue. The swelling gradually subsides and the pedicle is severed. Gillies closes the forehead scar and cuts away the excess tissue. He shapes the nostrils and defines the shape of the new nose, cutting or pulling in excess skin. By the time Spreckley is discharged his face is almost as good as new. The transformation is truly remarkable. Looking at him, you would never know that his nose had been rebuilt from his ribcage and forehead.

Spreckley was so grateful that he named his son Michael Gillies in honour of the surgeon who had restored his face.

The techniques Gillies had used for Spreckley were courageous, innovative and largely experimental. Although his operations were meticulous and his antiseptic techniques rigorous, there was always a risk that something could go wrong. The thing Gillies feared most was infection. If a wound became infected, there was little he could do.* Nevertheless, the cases that were coming to the hospital demanded that he try even more daring operations.

A STEP TOO FAR

Second Lieutenant Henry Lumley of the Royal Flying Corps was barely recognizable as human. His face was no longer covered in skin – it had melted into a red shiny mask of thin epithelium. His eyes were wide sockets with no eyelids or brows. His nose was pulled upwards, his lips – if they could even be called lips any more – were wide and inflamed, and his mouth scarred.

Lumley had never seen combat. During his first mission, in the summer of 1916, his plane crashed to the ground in a ball of flames. Pilots were not issued with parachutes, so when the fuel tank caught alight, Lumley was trapped in a fireball of petrol. His face, scalp, hands, fingers and legs were all severely burnt. Some areas of his head were protected by his helmet and scarf, but no one knew how he had managed to survive. He might well have been better off dead. Lumley was admitted to Queen's Hospital on 22 October 1917. He had spent the previous year being patched up by various

* *There were no effective antibiotics until the discovery of sulphonamide drugs in the 1930s. By the end of the Second World War, military surgeons also had penicillin at their disposal, which dramatically cut the number of hospital deaths.*

medical centres before he was finally referred to Gillies. It was the surgeon's toughest case yet.

Over the next month, Lumley was made as comfortable as possible while Gillies planned a series of operations. The surgeon proposed using skin from the pilot's chest to re-create his face. He would connect it with pedicles from the pilot's neck, and augment it with flaps of tissue from his shoulders. Gillies also decided to use paraffin wax and even attempt using a skin graft from another patient.

The first operation to prepare Lumley's face goes reasonably well and the patient seems to be making good progress. The second operation is about to begin. Lumley is anaesthetized on the operating table, his body propped up so his head is high. He has been stripped to the waist and his chest painted yellow with iodine. On the skin of his chest Gillies has drawn a face. There are spots for the eyes, marks for the nose and a long, narrow gap for the mouth. This outline will be Lumley's new face. It is a daring plan.

Gillies cuts and scrapes away the scar tissue from Lumley's face, leaving it horribly raw and red – blood seeping through to cover it in a glistening sheen. He then carefully cuts along the pencil lines on Lumley's chest until he has created a large (face-shaped) flap of skin. He lifts this up and places it across the airman's face, making sure to line up the holes for the eyes, nose and mouth. Then he begins to sew. Carefully and methodically, he attaches the new face across the remains of the old. When he has finished, he dresses the chest wound. The whole operation takes five hours. The surgeon is exhausted. The patient is terribly weak, his pulse faint. Now it is a question of waiting.

The first day after the operation, Lumley is definitely improving. The blood supply from the pedicles to the face seems to be working. On the second day, the graft starts to become infected. The doctors work desperately to stop the infection spreading. They try massaging the skin, pricking it with needles and cupping it (see Chapter 1) to increase the blood supply. By the third day, Lumley's new face is completely gangrenous. The pedicles from the shoulders are no longer supplying any blood and are gradually withering away.

By the tenth day, the dead skin has to be scraped off. Gillies records in the case notes that a foul discharge was expelled. The remaining pedicles are now only barely attached, and Gillies does what he can to save the blood supply. The doctors cleanse the wounds and spray them with paraffin wax. Later that day the patient is moved to an open-air hut in the hospital grounds.

Day fourteen and all the grafted skin has almost completely come away. For once, however, Gillies can report some good news. Lumley's chest seems to be healing and his face is no longer so infected. On 3 March Gillies starts a new treatment, using an ultraviolet lamp to encourage healing on the chest. By this time he has given up trying to save the face graft; he is now desperately trying to save the man's life.

Second Lieutenant Henry Lumley died of heart failure on 11 March 1918. Gillies had pushed plastic surgery to its limits, but with Lumley he realized that he had gone too far. Gillies wrote that his 'desire to obtain a perfect result somewhat overrode surgical judgement of the general condition of the patient'. He added, 'Never do today what can be honourably put off until tomorrow.'

Despite this terrible setback, Gillies achieved some fantastic advances in plastic surgery. Probably his greatest innovation was to adopt a Russian idea known as the tube pedicle. Instead of grafting exposed flat pedicles of skin which, as Lumley's case had proved, were prone to infection, he rolled the pedicle into a tube. This meant that all the delicate living tissue was enclosed within an outer layer of dead skin, providing it with a waterproof and infection-resistant cover.

But even the tube pedicle had its limitations. Skin could be moved only between adjacent sections of the body. A pedicle could be taken from the shoulder to the face, or the chest to the chin, for instance, but it was impossible to use the technique to take skin from the leg to the face unless the patient curled up in a ball for weeks on end. This made reconstructive surgery for a patient with burns across the whole upper body practically impossible.

As he was contemplating this problem, Gillies had a genuinely original idea; he called it the waltzing pedicle. What he would do was cut a pedicle from the leg and swing it upwards to attach it to the arm. Then, once the blood supply was established after a couple of weeks, he would cut the end still attached to the leg and swing it from the arm up to the face. By waltzing pedicles in stages to the site where they were needed he could safely take skin from anywhere on the body.

With the German push of 1918, more and more casualties were arriving at the hospital. Gillies worked all the hours he could while training up a new generation of plastic surgeons. Soon the wards were filled with patients covered in tubes of flesh; hoses of skin protruding from their legs, arms and faces; pedicles waltzing up their bodies.

Take the case of Private A.J. Sea, for instance, admitted to Queen's Hospital in June 1919. Since his injury, Sea had spent a year in military hospitals, but there was only so much the surgeons could do for him. In April 1918 he had been shot in the chin. The bullet had shattered his lower jaw, ripping away the floor of his mouth, taking the skin, bone and muscle with it. An ugly metal brace replaced his lower lip, keeping his jaw from falling apart. Sea's chin flopped uselessly, a few remaining teeth on his upper jaw stuck out at precarious angles. The twenty-three-year-old had to take all his sustenance through a straw. Like most patients who arrived at Sidcup, his eyes had the haunted look of a survivor who had endured more pain and suffering in a few months than anyone should experience in a lifetime.

The process to rebuild Sea's face was long and painful. The surgery was meticulously planned and the patient well prepared. The first operation was scheduled for August 1919, when a tube pedicle was cut from the soldier's chest and attached to his forearm. In October the end of the pedicle still attached to his chest was cut and attached to his missing chin, where it was held in position by straps. Six weeks later the surgeons took the end of the pedicle that was still attached to his arm and sutured it to his chin. The three operations were successful but, if anything, Sea's appearance was worse than ever. He now had a loop of skin passing beneath his mouth like a handle.

In March 1920 a large tube of skin was taken from his right shoulder. In September (more than a year after admission) a pedicle tube was taken from his neck, and work started to build the lining for the floor of Sea's new mouth. By December 1920 the private had undergone a total of ten operations, and in between he had received countless dressings, X-rays and examinations. By now Sea's chin was a dangling sack of skin covered in lines of stitches. Attached to it was a pedicle that passed around his neck and disappeared into the back of his shoulder. Another six operations followed over the next six months, until in August 1921 – two years after the surgery started – Private Sea was sent to a convalescent home to recover.

Sea was finally discharged from hospital in November 1922. His face was completely rebuilt. Although still disfigured, he had a mouth, jaw and lower chin. His broken teeth had been replaced by dentures, and his mouth had lips. Despite some scarring on his face and neck, he looked perfectly presentable. He would have only limited movement in his jaw, but at least he now *had* a jaw. Private Sea's life had been transformed. The last picture taken of him before he left the hospital even suggests that he was trying to smile.

In total, more than ten thousand operations were performed by the surgeons at Queen's Hospital. In all, only fifty men were lost – an incredible achievement given the ambition of the operations and the lack of antibiotics. Without surgery, many of the men might have survived, but with faces so damaged that their lives would have been a living hell. Gillies did his best to give his patients back their dignity.

Harold Gillies left Sidcup in 1919 to work on a definitive textbook of reconstructive surgery and set up a private practice. One of his first patients was recruited in an ethically dubious fashion while Gillies was staying at an inn during a fishing trip in Derbyshire. He noted that the daughter of the innkeeper was a 'comely lass' but she had a 'fearsome nose'. While he was out for the day, Gillies left a draft of his new book on the dressing table, open at the section about nose reconstruction. When he returned to London the girl contacted him and asked to be taken on as a patient. Gillies later admitted that it was a 'disgraceful' way of obtaining work, but at least the girl got a prettier nose.

Gillies was finally knighted for his services to surgery in 1930, although many people argued that the honour should have come years before. By that time he had accumulated piles of letters from grateful patients – from soldiers suffering with shattered jaws or burnt faces to children with harelips or cleft palates. Gillies' surgical skills had touched thousands of lives. He also had a reputation for kindness. He was known sometimes to waive the fee for those who could not afford to pay. The techniques he developed in Sidcup would be taken up by plastic surgeons around the world, and twenty years later would be adapted for a new conflict with even more terrible challenges.

McINDOE'S ARMY

Somewhere over England, 16 March 1944, 11.20 p.m.

Something had gone wrong and there was nothing the crew could do. The Wellington bomber was plummeting towards the ground. It dropped 300 feet in only a few seconds, then smashed into the earth, its tanks full of fuel. The explosion lit up the night sky and flames tore through the twin-engine plane. The Wellington's fuselage was covered in stretched fabric, and this burnt like paper, rapidly peeling away to reveal the metal skeleton underneath.

Nineteen-year-old navigator Bill Foxley forced open the plastic dome* on top of the aircraft and began to scramble to safety. Remarkably, he was hardly injured; an incredibly lucky escape.

Then Foxley heard the wireless operator's cry for help. He could hardly leave his friend to be cremated, trapped within the disintegrating airframe. Foxley lowered himself back through the hatch. The heat was unbearable, a violent wall of scorching flame. With the adrenalin pumping and the aircraft falling apart around him, Foxley hardly noticed that the skin on his hands was being seared on the smouldering metal struts, or that the flesh on his face was being stripped away by the heat. He reached his comrade and pulled him out. It was only when Foxley was well clear of the aircraft that he realized how badly he was now injured. His whole body seemed to be on fire.

He was admitted to the Queen Victoria Hospital in East Grinstead, some forty miles south of London. The Queen Victoria was the Second World War equivalent of Gillies' Queen's Hospital at Sidcup, and most severely burnt airmen ended up there. The men's recovery was overseen by Gillies' cousin, the brilliant and charismatic surgeon Archibald McIndoe. His job was to rebuild the airmen – ideally so that they could return to battle – but at the very least so that they could live a normal life after the war. With aviation fuel burning at temperatures of around 700°C, the surgeon faced an enormous challenge.

* *The dome or 'astrodome' was usually used for navigation. It enabled the navigator to see the stars, and he could use a sextant to fix the aircraft's position. The dome also doubled as an upper escape hatch.*

At the beginning of the war the majority of casualties had been airmen in Hurricanes and Spitfires defending the skies over southern England during the Battle of Britain. It was a horrendously dangerous occupation, and almost every day pilots would fail to return from their missions. The high-performance aircraft were packed with fuel, so if they were hit, the pilots had a good chance of being incinerated.

Both types of aircraft carried fuel tanks between the cockpit and the engine, but the Hurricane also had a 25-gallon tank in each wing. Unfortunately, a design flaw in the early Hurricanes meant that there was no fireproofing between the wing tanks and the cockpit. If a tank blew up, the cockpit became an oven surrounded by flame. Pilots were urged to keep the cockpit hood closed for as long as possible. Once they opened it, the flames tended to be drawn inwards. One airman described how he saw the dashboard melt and run like treacle before he was able to haul himself clear. Unlike in the First World War, at least these pilots had parachutes. However, bailing out into the English Channel was not a pleasant prospect because the icy salt water stung their wounds and hypothermia quickly took hold.

As the war progressed and the Allied raids on Germany intensified, more of the casualties were from bomber crews. There was a never-ending stream of new admissions to East Grinstead. Injuries ranged from shrapnel wounds to fuel burns. One patient was even admitted with frostbite. The rear door of his Lancaster had been blown open and his fingers had been frozen to the fuselage in his efforts to get it shut. To reconstruct these airmen, McIndoe had adapted and refined the techniques developed by Gillies during the First World War.

By 1944 the procedures were well established, the hospital well equipped and the staff well versed in caring for the victims of severe burns. Patients were immersed daily in specially designed saline baths to prevent infection and help their wounds to heal; new ways had been developed to deliver anaesthetics during the increasingly long and complex operations; and by the end of the war patients were being treated with penicillin. But, above all, McIndoe relied on the waltzing tube pedicle.

Ward Three of the East Grinstead hospital was bright and clean. There were fresh flowers on the tables, but nothing could disguise the nauseating smell of burnt flesh. Visitors, already desperately trying to cope with the visual onslaught, would frequently gag on the acrid stench. The beds were arranged in two long rows, and walking past them you could see the various stages of reconstructive surgery. Some patients were swathed in bandages, some had slings, but most had faces hung with pedicles – long hoses of skin that would soon be noses or jaws, lips or chins.

For the staff at East Grinstead, Bill Foxley was another typical case. Most of the skin on his face had been vaporized. It was distorted and ugly. His upper lip was burnt away, and the lower part of his nose had melted. It hung like dripping candle wax, leaving his nostrils flared upwards. His flesh was blistered and glistening, red and raw. His right eye was little more than a slit, blinded by the fire. His left was inflamed. Neither eye had brows or lashes. Worse were his hands. They resembled swollen gnarled stumps, the fingers fused together into a ball of flesh – a coagulated mass of tissue, bone and muscle all melded into one.

McIndoe's task was to rebuild Foxley's distorted face and do what he could for the airman's horribly damaged hands. Over the next few months Foxley had a series of operations to gradually restore his features. First, the surgeons took a tube of skin from his shoulder to his nose. Three weeks later it hung from his cheek to his nose giving his head the appearance of a jug. Finally, after a further three weeks, they used the tube to rebuild his upper lip. Nine weeks later the waltzing pedicle had done its job: Foxley's face had been successfully rebuilt.

The results of McIndoe's operations were even more impressive than Gillies' achievements during the First World War. Although Foxley's face was still somewhat distorted, in only a few weeks McIndoe had given him a new nose, lips and glass eye. The surgeons had also managed to separate what was left of his fingers and partly rebuild his hands.

But McIndoe was more than just a great surgeon; he was also a great psychologist. The patients at East Grinstead did not spend their days lying in bed; they were encouraged to get out and about. After all, most of these men were young and fit. Until their injuries, they had lived life to the full. Indeed, airmen were notorious for their fast living, and keeping them cooped up would do nothing to help them.

Most of their injuries were external and cosmetic, which meant that they were perfectly capable of moving around. So, with tube pedicles dangling from their faces, they could be found playing football in the grounds, drinking beer in the local pubs or watching films in the town cinema. McIndoe made great efforts to ensure that his patients were integrated as much as possible into the East Grinstead community. He encouraged local people to visit the hospital, and gave talks to explain the work that was done there. As a result of his efforts, the men were received as guests in local homes and were treated with respect in pubs and restaurants.

The town was proud to play host to the airmen and became part of their therapy – a stepping-stone between the hospital and the wider world. A world where perhaps they would not always be treated so well. During the First World War Gillies had found that, despite his best efforts, his discharged patients met with little public understanding. Their return to Civvy Street had sometimes been brutal. They were haunted by their looks and shunned by society. Many led isolated lives or ended up in dead-end jobs – selling matches or begging. One former patient even found employment as an 'elephant man' in a travelling circus. Many suffered from depression. Some committed suicide.

McIndoe wanted his patients to be treated as the heroes they were, not freaks to be locked up behind closed doors or laughed at in a circus. He encouraged the wounded men to support each other – to wear their injuries with pride. They called themselves the Guinea Pig Club, produced their own magazine, had a little emblem (a guinea pig with wings) and even their own anthem. This is the first verse:

> We are McIndoe's army, we are his guinea pigs.
> With dermatomes* and pedicles, glass eyes, false teeth and wigs.

* *A dermatome is a surgical instrument used to cut away slices of skin for grafting. In the song, however, it could equally be referring to the slices of skin themselves.*

A night out with the Guinea Pig Club could be a peculiar thing to witness. Their favourite haunt was the Whitehall restaurant in East Grinstead, where the manager, Bill Gardener, became almost as important to their rehabilitation as McIndoe himself. Gardener took a special interest in the men from Ward Three. He drank with them at the bar, but made sure they did not drink too much. He chatted with them, helped them laugh and managed to steer them away from moodiness or depression. Other places in town were similarly accommodating. Seats were reserved for the Guinea Pig Club at the cinema, and they were regular guests at local dances.

At any of these places you could see badly disfigured men, some with bandages, most with tubes of flesh hanging from their faces. They would be seen laughing, joking or chatting up the local girls. Some had only stumps for hands and needed help to drink. Their friends would lift the drinks to their lips and assist them later when they needed to go to the lavatory. Often these strange-looking men were accompanied by nurses from the hospital (whom they would also be chatting up). Gradually, the men of the Guinea Pig Club overcame their injuries and regained their dignity. Many of them married local girls, and very often local nurses.

Word of the hospital spread and soon gained national attention through newspaper and magazine articles. Britain's most popular entertainers came to visit the famous heroes of Ward Three and give performances in the town. Joyce Grenfell and Flanagan and Allen were among the stars to entertain the Guinea Pigs. The hospital was visited by senior politicians and military figures. The local paper reported that the hospital's work 'and the work of their splendid staff was known throughout the world'.

McIndoe's army was a triumph. The surgeon restored the faces of Allied airmen but, above all, he restored their pride. Bill Foxley is one of those proud survivors and attends the regular reunions of the Guinea Pig Club. Like many of McIndoe's patients, he looks back on his time at East Grinstead with affection. He recalls an occasion when the surgeon took a group of them into London. 'It frightened the life out of people,' he says, 'but that was all part of the game.'

Between them, Gillies and McIndoe had also restored the reputation of plastic surgery. They had developed new techniques and made tremendous advances in improving the appearance of their disfigured patients. But as 'beauty crank' Gladys Deacon had demonstrated, and as Gillies had discovered with the innkeeper's daughter, you didn't need to be badly injured to seek the advice of a plastic surgeon. There were lots of people who wanted to change their appearance, and for the gifted surgeon a whole new post-war world of opportunity was opening up.

A WHOLE NEW USE FOR A TUBE PEDICLE

London, 1946

Harold Gillies had been working as a government consultant during the war, but now he was ready to go back into full-time private practice. The rich and famous came to his house, just off Harley Street, for discreet facelifts, tucks or enhancements. He applied everything he had learnt from battle wounds to the fading faces of Knightsbridge and Mayfair, and was pulling in the equivalent of some £1.3 million a year. Plastic surgery had made his reputation, but cosmetic surgery was making him rich. Now aged almost seventy, he was about to perform an operation that would guarantee him a place in the history books.

Laura Maude Dillon was born the wrong sex. She spent her life convinced that she should have been a man. She dressed in men's clothing and could pass herself off as a man in the street. But this wasn't enough for Laura; she wanted to *be* a man. She was determined to transform herself physically into the opposite sex. When she made the decision in the late 1930s this was hardly an easy thing to do.

First, there were legal and social implications – how on earth would society treat her/him? Would she even be allowed to do it? There were other practical problems too. No surgeon had ever tried to turn a woman into a man before. But Laura was determined, and she managed to persuade a doctor to prescribe testosterone tablets. Her voice and appearance began to change and, during the war, she underwent a double mastectomy to remove her breasts. She was becoming androgynous, but was not yet a man. Finally, Laura was put in touch with Harold Gillies.

During a series of operations, carried out in the utmost secrecy, Gillies used his tube pedicle technique to build Laura a penis. First, he cut a tube of skin from her side and looped it around to her crotch. He then filled the tube with a frame of cartilage to give it bulk and structure. Once the blood flow was established, the end of the tube connected to her side was severed and the appendage gradually shaped into a penis. Finally, a rubber tube was connected to her urethra so that she could urinate through the new organ. Thanks to the tube pedicle, Laura became Michael. Gillies had successfully performed the world's first female to male sex change operation.

Michael's new penis was only cosmetic – he would never be able to achieve an erection, which meant he could never have a full sexual relationship. But being a man made it a lot easier to have a career. Michael enrolled in medical school under his new legal name and eventually qualified as a doctor. He even wrote a book, describing people who were born with the mind of one sex and the body of the other. No one reading the book guessed that he was actually describing himself. In fact, few people would ever have known that Michael was born Laura if it had not been for his aristocratic background.

Michael's brother was Sir Robert Dillon, the 8th Baronet of Lismullen. In *Debrett's* guide to the British aristocracy Michael was listed as Sir Robert's heir. However, in the rival publication, *Burke's Peerage*, Sir Robert's heir was given as Laura. The birth dates of Laura and Michael were the same, and it didn't take long for someone to realize that they were the same person. During research, the editor of *Debrett's* had come across the amended birth certificate that had transformed Laura into Michael.

The story broke in 1958 – and what a story it was. A sex scandal involving the aristocracy: what could be better? The world's press were all over it and set about tracking Michael down. They found him on a freighter in Philadelphia, where he was serving as the ship's medical officer. Reporters persuaded the reluctant doctor to give an interview. He certainly looked like a man. He was described as bearded and smoked a pipe. Dillon told the newsmen that he had been born suffering from hypospadias. This is a condition found in males where the opening of the urinary tract is not at the tip of the penis. Dillon had never in fact suffered from hypospadias; physically, he had been born a perfectly healthy girl, but was justifiably unwilling to give the reporters the truth. He said that he had undergone the operations to make him a more complete male.

He hated the attention and wanted to be left alone. Now the story was out, this seemed unlikely. So, rejected by society, isolated and depressed, Michael fled to India and eventually ended up in a Tibetan monastery, where he became a monk. He devoted the rest of his life to Buddhism and writing. Despite the prejudice he had encountered, he later wrote how he owed his life and happiness to Sir Harold Gillies.

When Gillies died in 1960 reconstructive surgery still relied on the tube pedicle. But the pedicle – even the waltzing pedicle – had its drawbacks. As it needed to be kept attached to its blood supply, moving tissue around the body took weeks. Patients had to endure straps or contraptions similar to those developed by Tagliacozzi to keep the pedicles in place, and suffer the awkwardness (and embarrassment) of having loops of flesh dangling around their bodies. There had to be a better way. Finally, by the 1970s, surgeons had come up with a solution: the operating microscope.

Today surgeons can take tissue from anywhere on the body. They use a large microscope positioned over the operating table to connect together minute blood vessels less than two millimetres across. Once the microscope is swung into place, they employ impossibly small needles and minute threads, narrower than a human hair, to make tiny, precise stitches. When Chinese surgeons first attempted microsurgery forty years ago, they unpicked a pair of stockings and used the fine nylon thread. Microsurgery is the same technology that made Clint Hallam's hand transplant possible (see Chapter 3).

What Gillies and McIndoe did not realize is that the transplanted tissue needs only a single artery and single vein to keep it alive. So even a relatively large swathe of tissue – skin, bone and muscle – taken from, for instance, the leg can be grafted on to a patient's face as long as it is connected by two blood vessels. Rebuilding a patient's jaw can be done in a single operation rather than over a period of months. Operating under the microscope has revolutionized reconstructive surgery and consigned the tube pedicle to history, although pedicles are still occasionally employed when all else fails.

But even the technology of microsurgery has its limits. As any before-and-after images of reconstructive facial surgery show, there is still a fundamental problem in repairing a face with tissue from other parts of the body. The difficulty is that the skin always looks like the area it has come from. The skin of an arm is different from the skin of a face – it can be darker or hairier – and when it is moved around the body this is all too apparent. Some surgeons believe the only solution is to transplant the skin from someone else's face.

In 2005 thirty-eight-year-old Isabelle Dinoire received a partial face transplant after being severely mauled by a dog. It was an incredible technical achievement for the French surgical team who carried out the operation, and now some surgeons are planning another huge step forward. They want to abandon traditional reconstructive surgery altogether and carry out a *full face* transplant. For some victims of facial disfigurement, this might be their only hope.

The story of Jacqueline Saburido illustrates the point. The bright, pretty twenty-year-old Venezuelan had moved to Austin, Texas, to study English. On the night of 18 September 1999 she was on her way home from a party, sitting in the front passenger seat of a car being driven by another student; three other friends were in the back. It was four in the morning, the road was dark. Suddenly an SUV veered across the carriageway towards them. Its driver was drunk.

When paramedics reached the scene, the front of the car Jacqui was travelling in was crumpled, the engine ablaze, broken glass across the road. The driver was dead – crushed by the steering wheel. One of the back seat passengers was also dead. The other two were pulled free, but Jacqui was pinned into her seat by the dashboard. She screamed for help as the flames reached higher. The paramedics tried to put out the fire but could do nothing to free her.

Then the screaming stopped.

When the firefighters arrived they doused the flames. Jacqui's flesh steamed as they gently turned the water on her body. Her seat had melted, the car interior was blackened by fire. Everyone looking at this awful scene of destruction assumed she was dead. It was a relief really, the screaming had been unbearable. Then Jacqui moved. She was still alive.

Almost two-thirds of her body was severely burnt. Her face was almost completely destroyed, her hair incinerated, her skin cracked and charred. Her hands had disintegrated into stumps, and she had several fractured bones. No one expected her to live for long. The driver of the SUV walked away from the crash, although he was later convicted and imprisoned for drink-driving.

With round-the-clock care in a specialist burns unit, Jacqui gradually started to recover. Since September 1999 she has undergone more than fifty operations. Surgeons have done their best to rebuild her face; they even managed to restore an eyelid that had melted in the blaze. But now they've reached the limits of traditional reconstructive surgery and there is little more they can do. Jacqui's face remains terribly disfigured. Her features are crumpled, her neck sagging, her skin a blotchy, crinkled patchwork. She has no hair, eyebrows or lashes. Her nose is flattened and distorted, her nostrils drawn upwards. She has only the remains of a single ear, and her left eye is swollen. Jacqui is still recovering from the events of 1999 and has devoted her life to campaigning against drink-driving.

After examining cases like Jacqui's, plastic surgeons such as Peter Butler believe that face transplants are the only way forward. Butler is one of Britain's top plastic surgeons and uses imaging techniques to simulate the effects of a face swap. On a computer screen his team can virtually place one face over another. In theory, the technical problems of a face transplant have already been overcome. The surgery is perfectly possible, although the cocktail of drugs to prevent rejection would probably take ten years off a patient's life. However, there are big ethical questions over whether it is right to take the face of one person and transplant it on to another. Our faces define us – how would a new face change us? And what about the donor family – how would they react to seeing the face of a loved one on somebody else's body?

Plastic surgery has come a long way since the brutal operations conducted in early India, or Tagliacozzi's leather corset and pedicle. The real triumph of plastic surgery has not been the cosmetic surgery for the 'beauty cranks' – the botox, the silicone or the face-lifts – but the effort that has gone into fixing terribly damaged faces.

Over the centuries surgery has restored the faces of syphilis victims, soldiers, airmen and the victims of fire or car crashes. Today surgeons can save the lives of even the most badly burnt and injured patients, such as Jacqueline Saburido, but despite all the advances in modern medicine they can only do so much.

Soon (possibly by the time this book is published) someone in the world will have received the first full-face transplant. You can bet the story will be a sensation. However, advances in tissue engineering will eventually enable surgeons to grow swathes of skin in the lab. They have already been able to grow a human ear on the back of a mouse. One day it might even be possible to construct an entire new face from samples of a patient's own DNA. We can only hope that the technology is used to reconstruct the faces of the victims of conflict or tragedy rather than to boost the vanity of ageing Hollywood stars, Page 3 models or the Gladys Deacons of this world.

CHAPTER 5
SURGERY OF THE SOUL

THE MAN WHO SHOULD HAVE DIED

Vermont, 13 September 1848

The navvies said that Phineas Gage was the best foreman they'd ever had. The twenty-five-year-old was fair and honest, a good worker and a fine leader. He was employed on the Rutland & Burlington Railroad. The promoters of the railroad hoped it would soon make them rich: winding through the wooded hills of New England, it would link Vermont to the cities of the east coast, bringing trade and creating new markets for the state's agricultural and mineral wealth. It was a fine enterprise indeed, and one well suited to an industrious and practical worker such as Gage.

Gage's work gang had been toiling since early morning near the town of Cavendish. They were building the roadbed – clearing and levelling the land in preparation for the rails. The plans called for a deep cutting, which had to be blasted through the granite hillside.

When it was finished, rock would dwarf the trains as they rounded the sharp bend in the track. It would be a proud moment, thought Gage, when the first steam engine – wheels pounding, head lamp blazing – rolled along the track into town. For that to happen there was still much work to do. Gage would make sure it was done, on time and to the highest standards.

As foreman, Gage was highly skilled in the use of explosives, but it was a tricky and dangerous job. First his men would drill a hole in the rock – using a manual drill on solid granite was not an easy task. Gage made sure that the hole was carefully positioned so that the natural fractures in the rock could be used to maximize the effects of the explosion. Next he lowered a measured amount of gunpowder into the shaft and inserted a fuse. He tamped the powder gently with his tamping iron before adding a layer of sand. The sand helped confine the explosion to a small space, focusing the charge into the rock rather than back up through the hole, which was simply a waste of good gunpowder. Finally, Gage tamped the sand good and hard, lit the fuse and stood well back. It was a job he did every day.

Gage was so practised with explosives that he even had his own custom-made tamping iron. The three foot seven inch-long iron bar was an inch and a quarter in diameter. The bar was round, flat at one end – the end he used to pack the explosives and sand – and tapered to a point at the other. It was more than a crowbar; it was styled almost like a javelin. A fine iron bar for an iron-willed man. An unfortunate description given what was about to occur.

It was half past four. They were nearing the end of another hard day and Gage could not wait to get back to the inn where he was staying. Most of his men were looking forward to an evening drink, but Gage rarely touched alcohol himself. They were loading lumps of rock on to a flat car as Gage prepared to blast another section from the hillside.

He lowers the string of the fuse into the hole and pours in the gunpowder. He begins to tap the powder gently with his iron. Distracted by the work going on behind him, he leans forward over the hole. Perhaps he forgets that the sand hasn't yet been poured, or perhaps he slips. But when he tamps the iron again, it goes in too hard and catches on the granite. It ignites a spark.

The gunpowder explodes.

The iron rod shoots out of the hole like a bullet from a gun. It goes straight through Gage's cheek, passes through the floor of his left eye into the front of his brain and tears out of the top of his head. His skull is splayed apart as the iron continues its journey upwards, eventually returning to earth some eighty feet away, smeared with blood and bits of brain. Some of Gage's brain is later found splattered across the rocks where the rod landed.*

Gage was knocked on to his back by the force of the explosion. His men ran across to find him twitching on the ground. A few moments later he spoke. Then, to everyone's astonishment, he got up and started to walk towards the road. He was helped on to an ox cart and driven the three-quarters of a mile into the centre of town. When he arrived at the tavern of Joseph Adams, where he was lodging, he walked with only a little assistance and sat in a chair

* *The men who found the iron reported that it was 'covered with blood and brains'. They washed it in a nearby brook, but it still had a 'greasy' appearance and was 'greasy to the touch'.*

on the veranda. He chatted with some of the men who gathered around and answered questions about his injury. Gage had rarely missed a day's work in his life and said he was keen to get back to the railroad.

When Dr Edward Williams arrived at around five o'clock he could not believe what he seeing. It made no sense – how could this man possibly be alive? Gage remained perfectly lucid, insisting that the bar did indeed pass right through his head. One of the labourers corroborated the story: 'Sure it was so, sir, for the bar is lying in the road below, all blood and brains.'

Despite the burn marks on Gage's cheek, the copious amounts of blood dribbling down the poor man's face and the fragments of bone sticking from his head, Williams was still unable to accept what had happened. It wasn't until Gage started vomiting a large quantity of blood and, as Williams noted, 'about half a teacupful of the brain, which fell upon the floor' that the doctor finally came round to the idea that Gage had survived the firing of an inch and a quarter-wide tamping iron through his head.

Williams was completely flummoxed by the case, and seemed reluctant to administer any treatment. So an hour later, when Dr John Harlow arrived, Gage was still sitting on the veranda answering questions and recounting his dramatic tale; also occasionally vomiting blood, bone and lumps of brain that had dropped through the hole from the top of his head into his mouth. Harlow was impressed with how Gage 'bore his sufferings with the most heroic firmness'. Despite becoming increasingly exhausted from the massive loss of blood, Gage recognized the doctor at once and needed little assistance to make his way up the stairs to his room.

Harlow was much more practical although, unsurprisingly, somewhat taken aback by the mess. 'His person and the bed on which he was laid were literally one gore of blood,' he recalled. However, this didn't stop the doctor passing his fingers completely through the hole. 'I passed in the index finger its whole length, without the least resistance, in the direction of the wound in the cheek, which received the other finger in like manner,' he later reported.

Together the doctors cleaned and dressed Gage's wounds. They shaved his scalp and removed a few bits of bone and a stray piece of brain that was 'hung by a pedicle', as well as bandaging the burns on his hands and arms. Harlow pressed the jigsaw of bones on the top of Gage's skull back into position as best he could and left the man propped up in bed, where his bandages gradually became saturated with blood. A couple of the men volunteered to watch over him.

When Harlow returned at seven the next morning, Gage was still conscious. He had even managed to snatch some sleep during the night. Harlow didn't expect him to live for much longer, and the undertaker was called so that Gage could be measured for his coffin. It seemed the prudent thing to do. As the undertaker took his measurements, Gage's mother arrived to say her last goodbyes.

By 15 September Gage's condition had indeed deteriorated. He was passing in and out of consciousness, he was delirious and incoherent. On the 16th Harlow replaced the dressings but described 'a fetid sero-purulent discharge, with particles of brain intermingled'. That couldn't be good.

Harlow continued to visit his patient every day, and by the 22nd it seemed that the stubborn (and iron-willed) Gage was finally ready to die. He was hardly sleeping at all; he threw his arms and legs about as if he was trying to get out of bed. His body was hot, his wounds fetid. He even told the doctor, 'I shall not live long so.'

One month later Gage was walking up and down the stairs, even into the street. His wounds were healing rapidly and he was eating well. His bowels were described as 'regular' and he had even stopped vomiting globules of brain. By the end of November all the pain had subsided and Gage told the doctor that he was 'feeling better in every respect'. He could walk, talk and eat. There was only one problem: Gage was no longer Gage.

He described it as a 'queer feeling'. Others said the man had completely changed. The accident had radically transformed his personality. The railroad foreman who had once been described as sober, patient and industrious was now vulgar, impatient and impulsive. Gage was rude, they said, and could suddenly break forth into vile profanity. When he reapplied for his position as foreman his employers said the change in his mind was so marked that they refused to take him on. He was described as childlike in his attitude, but 'with the animal passions of a strong man'.

Gage's accident went beyond mere medical curiosity. When the iron bar tore away part of his brain it revealed the inner workings of the mind. It demonstrated that the brain is not some homogeneous grey pudding, but is made up of different parts doing different things. This is a concept known as localization, and would become vitally important for our understanding of the brain and for the first tentative advances in brain surgery.

Most of our personality, our sense of 'self', is contained behind the forehead, in the frontal lobes of the brain. These were the parts that were blasted away by Gage's tamping iron – the parts sprayed on to the rock or those that he later vomited across the floor. The frontal lobes are where we think and plan things. When the rod ripped through Gage's brain it tore away his personality and made him more impulsive. A century later surgeons would employ smaller rods to do much the same thing.

Gage never did return to the railroad. With his tamping iron as his constant companion, he travelled across New England. He eventually ended up in New York, where for a while it is said he became a sideshow in the famous Barnum's American Museum. For a few cents, punters could see a living man with a hole in his head. Although anyone expecting to see something truly gruesome would have been sorely disappointed. They could see (and perhaps if they were lucky, touch) the tamping iron, but the hole was now healed and there was little to show for Gage's trauma. Instead visitors could listen to Gage as he used another skull to regale his dramatic story.

In December 1848 Harlow's account of the case was printed in the *Boston Medical and Surgical Journal*. It was greeted with scepticism by the medical establishment, most of whom believed Gage's survival to be completely impossible. Surely Harlow must be mistaken? What would a rural doctor like Harlow know about the anatomy of the brain anyway? However, by 1849 Gage's case had attracted the attention of the new professor of surgery at Harvard Medical School, Henry Jacob Bigelow, who compiled a detailed account of the accident and paraded Gage (and his tamping iron) in front of surgical colleagues, suggesting that this was 'the most remarkable history of injury to the brain which has been recorded'. Thanks to Bigelow, Gage's accident would become a medical sensation and one of the most curious incidents in the whole history of surgery.

In the first days after the explosion it had been reported in the local paper as merely a 'Horrible Accident'. Workers died all the time on the railroad; it was hardly big news. But now, as more and more newspapers heard about the case, Gage's fame spread. He could have made a comfortable living on the medical freak show circuit – travelling around the USA from circus to surgical symposium (they often amounted to the same thing). He would be the nineteenth-century equivalent of a daytime chat show guest.

However, Gage's new impulsive nature took him in a different direction. The new Gage discovered that he enjoyed working with animals and went to work at a livery stable. For a while he cared for horses and drove a stagecoach in Chile. But with his health failing, he returned to the United States in 1859, finding employment on a farm in California. Then, in 1860, the accident that should have killed him finally did.

In February 1860, while ploughing a field, he suffered an epileptic fit. During these final few months of his life he started to suffer more and more fits and convulsions. Doctors did the only thing they knew how – which was to bleed him – but the treatment seemed to have little effect. Phineas Gage finally died in May 1860, twelve years after a three foot seven inch-long iron rod passed through his brain.

Although Harlow's treatment of Gage was exemplary, it is one thing to piece back together a fractured skull or even care for major head injuries such as those sustained by Gage, but it is quite another to open up the head and poke around in the brain – to have a crack at brain surgery. As far as most surgeons were concerned, any attempt to go further than repairing a head injury was to be avoided at all costs. Anaesthetics, advances in anatomy and, later, Joseph Lister's antiseptic operating techniques (see Chapter 1) might have transformed nineteenth-century medicine, but the brain was still a mystery, locked away in the sealed casket of the skull. Few surgeons were prepared to open this casket, and those who did usually came quickly to regret it. The only exception to the unwritten 'no operating on the brain' rule was the ancient practice of trepanning.

Trepanning is arguably the world's oldest surgical practice – although amputation is likely to run it a close second. It involves drilling a hole, from half an inch to two inches across, into the skull. The patient would have had their hair and skin scraped away before the prehistoric equivalent of a surgeon started to bore into their head with a sharpened stone or, later, a crude metal drill.

The practice of trepanning was in widespread use from around 10,000 BC, before the invention of reading or writing. The incredible thing is that archaeologists have found skulls with holes drilled in them all over the world. The evidence suggests that trepanning was being practised by many different peoples in completely different locations. These were communities that were segregated by geography. They had no possible way of contacting each other or, indeed, any knowledge of the others' existence. This implies that either many groups developed trepanning separately, or that the practice was passed down from our earliest human ancestors.

The big question is why? Why on earth would you want to drill a hole in someone's head? There could be any number of reasons, which historians can only guess at. In some civilizations, those who were trepanned also show evidence of head injuries, suggesting that trepanning was used as a treatment. It might also have been used to cure headaches, epilepsy or insanity. Perhaps it allowed demons to escape. There is speculation that it might even have given the recipient magical powers – a window, perhaps, to the gods.*

Back in the nineteenth century there were three major challenges facing budding brain surgeons: the risk of infection, the danger of haemorrhage and the fact that they had very little idea what each bit of the brain did. The issue of infection applied to any major surgery, but infection from operations was gradually being defeated as Lister's reforms were adopted. As for controlling blood loss, this was still a significant problem. The brain has more than four hundred miles of blood vessels and consumes nearly two pints of blood every minute. The scalp, brain and bone are all extremely bloody, so cutting into the skin, skull and membranes surrounding the brain means there is a good chance of a patient bleeding to death on the operating table.

As for understanding how the brain works and the functions of its different areas, these are things that scientists are still grappling with today. Phineas Gage gave surgeons an insight into the role of the frontal lobes, but any sort of accurate map of the mind was still a long way off. All this left surgeons powerless to help people with

* *You might think you need trepanning like a hole in the head, but it is still practised today. Surgeons have to use drills to access the brain, but there are also alternative therapy groups that recommend trepanning for all sorts of mental health conditions. In 2000, for instance, a British woman decided to do her own DIY brain surgery. Twenty-nine-year-old Heather Perry from Gloucester injected herself with a local anaesthetic before drilling a one-inch hole in her own skull. Unfortunately, the drill went in too far, damaging a membrane and requiring emergency medical help. Despite the mishap, she told reporters she had no regrets about the procedure.*

brain injuries or tumours. Or to find a surgical cure for insanity. Even so, despite limited knowledge and unrefined techniques, some surgeons were still prepared to have a go.

AN OPERATION ON THE BRAIN

Hospital for Epilepsy and Paralysis, London, 1884

Henderson's problems started in about 1880. The Scottish farmer was working in Canada when a piece of timber fell from a house and struck him on the head. The impact knocked him unconscious, but he recovered well and, apart from the occasional headache, returned to health. A year or so later he found that the left side of his mouth had developed a twitch. There was a similar sensation on the left side of his tongue. Within a few months, he was experiencing fits. They began with a 'peculiar feeling in the left side of his face and tongue' before spreading down the left side of his body. They culminated in convulsions and eventually loss of consciousness.

Henderson told the doctors that his symptoms had gradually worsened. He explained how he started to experience the twitching sensation on the left side of his face on a daily basis. The seizures increased in frequency until he was blacking out at least once a month. The twitching spread to his left hand and arm. The limb had weakened until he could no longer move it at all. By August he was unable to use his tools and was forced to give up work. By the autumn the paralysis had spread to his leg. He walked with a noticeable limp. Henderson was admitted to hospital on 3 November, but his condition was deteriorating by the day.

When physician A. Hughes Bennett examined Henderson, the case notes made depressing reading. Bennett had little doubt that a 'fatal termination was not far distant'. Henderson was keeping the rest of the ward awake with his screams; terrible cries from the violent, stabbing pains in his head. The headaches lasted for up to twelve hours at a time. He experienced seizures, attacks of sporadic twitching, violent tremors and uncontrollable vomiting. Bennett prescribed morphine for the pain, but no amount of ice packs or drugs seemed to give poor Henderson any relief. The situation was desperate; a 'fatal termination' seemed inevitable. But Bennett had one final trick up his sleeve.

There was no outward sign on Henderson's skull where the problem might be. Nevertheless, Bennett had studied localization and, without the aid of imaging equipment (no one had yet invented any), the physician diagnosed that Henderson was suffering from a brain tumour. What's more, he was confident he knew where it was: on the right side of the Scotsman's brain – the part that doctors believed controlled movement in the left side of the body.

Bennett decided that the tumour had to be removed, but as he wasn't a surgeon himself he enlisted the help of Rickman Godlee. A nephew of Lister, Godlee was well versed in the latest antiseptic operating techniques. Together he and Bennett planned the first operation to remove a tumour from a living human brain. The procedure would take place on 25 November 1884.

The operating theatre was prepared in the strictest accordance with Lister's methods. The instruments were soaked in carbolic; so too were the bandages and the surgeon's hands. Henderson was carried from the ward and laid out on the operating table, his head propped up on a wooden block. When everyone was ready, a gauze containing chloroform was placed over the patient's face. He was instructed to take deep breaths and gradually slipped into unconsciousness. An assistant started up the carbolic pump and soon a fine mist of antiseptic acid engulfed the area around the patient. They were ready to start.

To work out where to cut, Bennett had drawn a series of lines across Henderson's scalp. It was similar to using triangulation to obtain the position of a location on a map. He had tried to estimate where in his patient's brain the tumour was most likely to be. There were four lines in total and X marked the spot. Bennett indicated to Godlee where to make the first hole.

The drill squeals as Godlee cranks the handle and the bit grinds through the skull, becoming clogged with skin and fragments of bone. He makes sure to apply enough pressure to create the hole, but not too much in case the tool suddenly plunges inwards and gouges the brain. Godlee carefully removes an inch-wide circle of scalp and peers into the hole. An assistant holds an oil lamp over Henderson's head so that they can all get a better look. So far, so good. The outer membranes covering the brain – the meninges – look normal, but when Godlee sticks his knife through them, the brain pulsating beneath appears to bulge.

The doctors decide to proceed with the next hole. Godlee pushes the drill against the skull so that it is slightly overlapping the first hole, and begins to turn the handle. When he has finished drilling, he takes a hammer and chisel and starts to chip away at the jagged corner between the two holes. They can see more of the brain, but, after a quick discussion, they decide to make a third hole.

Once Godlee has finished it off with the chisel, they are left with a triangular aperture in the man's head.

Working slowly, Godlee starts to slice through the first layer of the membrane – the dura. He is careful to avoid a large blood vessel. When he lifts the surface of the membrane he can clearly see a transparent solid globule of tissue underneath. He has found the tumour – exactly where Bennett had predicted. Pulling apart the membrane a little further, he is able to wedge a narrow steel spatula between the tumour and the surrounding brain tissue. He slips his finger underneath to try to pull the mass free. He pulls too hard because the upper part of the tumour breaks open.

The operation is getting messy. Blood is oozing out over everything. As soon as Godlee mops it up with a sponge, the triangular opening in Henderson's head wells up again. Struggling to see what he is doing, the surgeon dips in a spoon and begins to scrape away at the remains of the tumour, trying as hard as possible not to remove too much healthy brain in the process. Removing the tumour leaves a hole around one and a half inches deep or, as Bennett puts it, 'a size into which a pigeon's egg would fit'. Later, when they have cleaned it up, they will find the tumour to be 'about the size of a walnut'.

Hands covered in blood, everything else now totally soaked in stinging carbolic from the spray, Godlee starts to close the wound. To do this he employs another recent surgical innovation (a variation of which is still used today): an electrocautery. This is an advance on the old-fashioned cauterizing iron, which has done so much damage during amputations (see Chapter 1). Godlee inserts an electrode into the wound and holds it against the bloody tissue as his assistant throws a switch. The flesh sizzles and the bleeding slows. Satisfied, Godlee stitches together the dura, slipping in a rubber tube to drain any excess fluid, and dresses the wound in gauze. A mixture of blood and spinal fluid drizzles from the tube.

The whole operation has taken two hours. Henderson has remained unconscious throughout, but when he awakes he seems to have suffered no ill effects. Better still, the pain in his head, the convulsions and the twitches have all disappeared. His left side is still partially paralysed, but this is only to be expected. It looks as if Bennett and Godlee have done it. Henderson is cured.

Unfortunately, Henderson did not live long enough to appreciate this remarkable new surgical treatment. Despite Godlee's best efforts, the wound somehow became infected. Bennett speculated that this might have been as a result of the cauterizing apparatus or the sponges (or it could have been because the surgeons had not worn masks or gloves), but once the infection had taken hold, there was little the doctors could do. One month after the operation Henderson, like so many experimental patients in the history of surgery, was dead.

Whether an operation that ultimately leads to the patient's death can be described as successful is debatable. Bennett and Godlee's achievement was nevertheless considerable. They had done everything they could think of to prevent infection, and the technology they used – from the chloroform anaesthetic to the carbolic spray and electrocautery – was Victorian state-of-the-art. Bennett had accurately diagnosed a brain tumour, had identified exactly where it would be, and Godlee had managed to remove it successfully without the patient dying on the operating table.

Given that without the operation Henderson would certainly have died in terrible pain, Bennett and Godlee were probably right, on balance, to go ahead and deserved the acclaim they received.* They had made a major advance in neurosurgery, proving that it was possible to open the sealed casket of the skull and operate on the brain. Now it seemed that every other surgeon wanted to have a go.

Over the next twenty years, thousands of operations were carried out on the brain. In the United States alone more than five hundred surgeons attempted brain surgery between 1886 and 1896. These were all general surgeons who applied the same techniques to excising a brain tumour as they might to removing a diseased appendix. Like Bennett and Godlee, they would categorize operations as successes even though their patients subsequently died. The surgeons consoled themselves with the knowledge that their patients would have died anyway; but this didn't stop them pocketing a healthy fee for the operation.

In 1889 the German surgeon Ernst von Bergmann compiled a review of the mortality rates from brain operations. His study made depressing reading. On average, half the patients undergoing brain surgery died. Some bled to death on the operating table after surgeons accidentally severed a major blood vessel, sending a shower of blood spurting from the wound. Other surgeons managed to remove tumours successfully only to find that they couldn't shove the brain back in again. Lobes of brain tissue would

* *Of course, as it turned out, Henderson* did *die in terrible pain, only it was from meningitis rather than a brain tumour.*

bulge accusingly through the hole in the patient's head. Struggling to force it back in, they would find they could no longer draw together the flaps of dura or get the skull back on. It was like trying to close the lid of an overfilled suitcase, and would almost have appeared comical had it not invariably ended with the patient's death.

If the surgical procedures themselves left a lot to be desired, so did the diagnosis and aftercare. Bennett had got the position of Henderson's tumour absolutely spot on, but other surgeons were not so lucky. The anatomy of the brain was only broadly understood. Surgeons would anaesthetize the patient, drill into their skull and cut into the membranes only to find a perfectly healthy brain underneath – thereby incurring all the risks of surgery without any hope of success.

One of the greatest killers, however, was infection – a problem that had been overcome in most general surgery. Time and again surgeons would operate, remove a tumour and successfully close the wound, only to have the patient die from infection a few weeks later. Even those who, like Bennett and Godlee, employed the very latest antiseptic techniques still seemed to come unstuck at this final hurdle. Soon even the most gung-ho surgeons decided that brain surgery was more trouble than it was worth and gave up neurosurgery altogether. The mortality rate was doing nothing for their reputation. Brain surgery remained in the Dark Ages. It desperately needed someone to make it safe.

THE MAN WITH ONE THOUSAND BRAINS

Peter Bent Brigham Hospital, Boston, 1931

Harvey Cushing was a god among surgeons. And he would often behave like one. Worshipped and feared in equal measure, his patients adored him while his assistants were terrified of him. Cushing was cold to his family and a bully to his friends, but a model of care and tenderness with his patients. Colleagues described him as hard and selfish. He was so focused on his work that when he was told his son had died in a car accident, he carried on with a scheduled operation anyway. When it came to brain surgery, Cushing was a miracle worker – the first true neurosurgeon.

A Cushing operation was an intense affair that could last for anything up to eight hours. He sometimes had another surgeon perform the opening of the skull and the closure at the end, but there was no doubt as to who was in charge. Cushing sat on a stool beside the operating table so that he was level with the patient's head. He worked slowly, methodically, pedantically. Every blood vessel was clamped off until the hole in the patient's scalp was surrounded by dozens of scissor-like clamps. He inserted smaller wire clips and painstakingly cut, scraped and cauterized as he removed tumours. In some cases these growths were massive – a witness reported one to be as 'big as an orange'.

Cushing was a tyrant in the operating theatre. He cursed his assistants if they failed to second-guess his every move, and barked at nurses if the right instrument wasn't placed in his gloved hand. He ordered surgeons out of the room if he thought they were being clumsy, and belittled his colleagues – usually in their presence. He demanded the same high standards from his staff that he expected from himself. But his results were exceptional. Only around one in ten of his patients died. Given that many were seriously ill and that antibiotics had not yet been invented, it was an impressive record.

On 15 April 1931 Cushing carried out his two thousandth tumour operation. His patient was thirty-one-year-old Ida Herskowitz. She had been suffering from debilitating headaches and was rapidly losing her sight. It wasn't a particularly complex operation (in relative terms), and the surgeon managed to remove a tumour successfully and restore Herskowitz's vision.*

Cushing had first become interested in operating on the brain when he was qualifying as a surgeon in the late 1890s. Despite the terrifying mortality rates associated with brain surgery, he decided that neurosurgery was going to be the next great surgical revolution, and he wanted to be part of it. Indeed, not only part of it – he wanted to lead it. With single-minded determination, he achieved his goal within a few years, and by the 1930s was at the height of his powers. Most of his innovations were relatively small, but together they made brain surgery effective and a good deal safer.

One of Bennett and Godlee's biggest problems had been the amount of blood that sloshed around as they were working. Cushing's first goal was to work out a way of stemming blood flow during an operation. He wanted to see what he was doing while

* *Ida Herskowitz was still alive thirty years later. Even though Cushing sometimes treated his staff abysmally, they were immensely loyal, and on completion of this landmark operation they presented him with a silver cigarette case and an elaborate celebratory cake.*

preventing his patients from bleeding to death. His answer was to make small clips from pieces of household wire and clamp them across arteries and veins. He also adapted a pneumatic cuff, originally designed for measuring blood pressure, to act as a tourniquet and reduce blood flow to the scalp.

Cushing was quick to adopt new technology. He was one of the first surgeons to use X-rays for diagnosis, and pioneered the use of an 'electric scalpel'. This device was an advance on the primitive electrocautery probe used by Godlee and Bennett, and allowed the surgeon to cut and seal tissue at the same time. Unfortunately, the electric scalpel could also burn and shock – both the staff and patient – and in one case sent a patient jumping, in the words of a witness 'like a frog', almost off the operating table. Still, when it worked the electric scalpel was a major improvement for controlling bleeding, and particularly useful for excising tumours.

The risk of infection remained a major concern for surgeons, and Cushing operated in conditions of the strictest cleanliness. Everyone in the theatre wore masks and the surgeon operated with gloves. He also appreciated the importance of aftercare. Following operations, patients were nursed around the clock by staff specially trained in dealing with neurosurgery cases. Sometimes patients were even kept in the operating theatre to keep the risk of infection to a minimum. This post-operative treatment was the forerunner of the intensive care units found in modern hospitals.

If Cushing gave every appearance of remaining emotionally detached from his family and colleagues, quite the opposite was true when it came to his patients. It is said the only time he talked of his son's death with any emotion was when he was comforting the parents of a dead child. There is a picture of him holding the hand of a man suffering from acromegaly, a condition caused by an overproduction of growth hormone from the pituitary gland, which results in an abnormal increase in height. Another touching photograph shows him holding a cuddly toy at the bedside of a poorly child whose head is swathed in bandages.

Cushing could not bear to let a patient die, and would do anything he could to help them. Patients spoke of how gentle and kind he was, and told of his sympathy and understanding. Unlike some of the other god-like surgeons around at the time (and since), he was not above helping to clean a patient or deal with their bedpan. And in return for this great care and his undoubted surgical skills, his patients bequeathed him their brains.

The Cushing Tumour Registry comprises a unique collection of photographs, notes, hospital records and brains. Lots and lots of brains. There are around one thousand of them in the Yale archives, collected over more than thirty years. They are arranged on shelves like jars of sweets. Each jar, labelled with details of the case, contains the disembodied brain of one of Cushing's patients. Each one is preserved in fluid, its folds and ridges helping to form a unique record of brain disease.

Cushing's legacy is represented by these jars, but is preserved by the techniques he developed – techniques that are still being used today. He helped train a new generation of neurosurgeons and his work led to future advances in neurosurgery. More than anyone, Cushing made modern brain surgery possible. Now surgeons could operate on brains with every confidence that their patients would survive.

Unfortunately, while Cushing was pushing forward the barriers of modern medicine, others seemed hell-bent on returning it to the Dark Ages.

WALTER FREEMAN, LOBOTOMIST

Washington DC, 1936

There were many reasons why Walter Freeman did what he did. The reasons were lying in the squalid wards of the mental hospitals, staring at the walls, screaming or moaning. The reasons were shouting at invisible demons or lying curled up in the corner of a rubber-walled cell. The patients of mental hospitals were Freeman's reasons; people with no hope.

In 1924, when Walter Freeman was first appointed as laboratory director at St Elizabeths Hospital in Washington DC, he was shocked by what he saw. When he strode through the overcrowded wards of the vast institution he felt a mixture of fear, disgust and shame. Fear of the patients who crowded around him, disgust at the excrement smeared on the walls, and shame that the doctors were powerless to do anything to help these poor people.

Psychiatric hospitals were known as snake pits. They were warehouses where society dumped the mentally ill; locked people away – often for years, sometimes for a lifetime. They were places of horror and hopelessness. Wards were packed with beds, with hardly any space to move between them. The sheets would be soiled, many of the patients ignored. There were too few staff, and many of those acted more like prison warders than hospital carers.

As he walked through the wards, past the padded cells and through the heavy steel doors, Freeman saw things that would make a lasting impression on anyone. There were young men squirming on the floor, their hands tied so that they could no longer claw at their skin. He saw patients being force-fed, their jaws clamped open by burly orderlies. Some patients would suddenly become violent and abusive, only to be dragged off to a cell. Others would be sitting, just sitting, staring into nothingness, as if their brain had simply shut down.

Although St Elizabeths was one of America's largest mental institutions, it was typical of others around the world. Admissions to psychiatric hospitals were growing by some 80 per cent each year, but the worst thing was that they could offer little in the way of treatment. For the most part, the best the staff could do was keep the patients alive. Those who attempted suicide were restrained or constantly monitored. The only hope was that the mentally ill would recover spontaneously after their period of 'rest' in the hospital. For most patients the stay in hospital had the opposite effect and their condition simply deteriorated.

By the 1940s what treatments there were relied on shocking the brain back to health – sometimes quite literally. Doctors would overdose their schizophrenic patients with injections of insulin to induce convulsions. Others preferred to use a drug called Metrazol to induce seizures. Metrazol convulsions were so violent that patients were contorted in agony, and many suffered fractures to their spine. Patients begged doctors not to put them through this torture but, as some doctors reported that the seizures were resulting in dramatic cures, their pleas usually fell on deaf ears.

The most controversial of all the shock therapies was ECT – electroconvulsive shock therapy. Invented by an Italian who had seen electric shocks used to stun pigs prior to slaughter, it appealed to psychiatrists because it was quick, cheap and easy to use, and was much more controllable than Metrazol. The other advantage of ECT was that it could be used to control the behaviour of patients. There is plenty of good evidence that ECT is effective at treating mental illness, and it is still used today under controlled conditions and with the full consent of the patient. However, in the 1940s, as ECT spread to hospitals across the world, it was quickly adopted as a way of keeping patients subdued.

The procedure was simple enough. Patients would be held down on a bed while electrodes were placed on either side of their head. Some ECT machines employed a Y-shaped electrode, like a catapult, that could be held by the doctor. When the current was turned on the electricity induced a seizure, leaving the victim passive and quiet. Aggressive patients could be given several shocks a day to keep them under control. Patients would be threatened with ECT if they did not behave.

This is the world Dr Walter Freeman was working in – a world he was determined to change. The theory of localization was now widely accepted and Freeman was convinced that mental illness was a result of a physical defect in the brain. It was a view backed by the apparent effectiveness of shock therapies. But rather than fire jolts of electricity through the brain, he wanted to change the whole way it was wired up.

Freeman was determined to get to the root cause of mental illness. In his laboratory he worked tirelessly, examining thousands of brains – slicing them, dissecting them. Day and night he measured the brains of dead mental patients and compared them with 'healthy' brains. Freeman was becoming an expert in brain anatomy, but however much he sliced, diced, measured and dissected, he could find nothing to distinguish the brain of a severely mentally ill patient from the brain of anyone else. It seemed like he had reached a dead end. He had wasted years of his life in pursuit of a physical defect that didn't exist. Then he came across the work of Portuguese surgeon Egas Moniz.

In 1935 Moniz had carried out a radical new operation. He called it a leucotomy. The procedure involved drilling several holes in the front of the patient's skull above the frontal lobes of the brain. Moniz then inserted an instrument he had devised, known as a leucotome. The device acted like an apple corer. When the surgeon pressed down on a plunger and rotated the leucotome, he could extract a brain core one centimetre wide. Usually, he would take about four cores of brain during an operation. Moniz could claim, with some justification, that around one-third of his operations were successful. He never came up with a scientific explanation for why leucotomies worked, but said they made his patients calmer and less agitated; they removed many of the symptoms of anxiety and psychosis. He believed the procedure had no effect on the intelligence of the patients and that it enabled them to lead normal lives once again.*

* *In 1949 Moniz received a Nobel prize for his 'discovery of the therapeutic value of leucotomy in certain psychoses'. There have been campaigns to have the prize posthumously taken away from him.*

When Freeman learnt of Moniz's research he became wild with excitement. Moniz had proved what Freeman believed all along: surgery was the answer. Freeman became convinced that many types of mental illness were caused by the connections between the thalamus – a small structure deep in the brain – and the frontal lobe. The thalamus, he believed, was the seat of human emotions. If he could only sever the connections in the front of the brain, it would dampen down all these terrible emotions and his patients would be cured.

Freeman became the Portuguese surgeon's biggest fan. He decided that he would adopt Moniz's operation and make it his. He could finally help the patients of St Elizabeths. It could make him famous – his name would be cited alongside other great medical pioneers. He could imagine it now: Walter Freeman – the inventor of the lobotomy.

But Freeman had a problem: he was not a surgeon. So he enlisted the assistance of someone who was – a young neurosurgeon named James Watts. Together they planned the first of Freeman's new lobotomy operations on sixty-three-year-old Alice Hammatt. The woman had been suffering from depression, anxiety and insomnia. She was sometimes suicidal, invariably agitated. Without treatment she would end up being admitted to a mental institution, where she would undoubtedly spend the rest of her life. To Freeman she seemed like the ideal patient. On 14 September 1936 in an operating theatre at George Washington University Hospital in Washington DC, Hammatt was put to sleep.

With Freeman directing the operation from a stool a few feet away, Watts cautiously cut three incisions into Hammatt's shaved scalp. Next he drilled a hole on both sides of her skull – above her ears and behind her forehead – and stuck the leucotome into the left-hand hole. Watts pressed the plunger on the instrument and cut the first core of brain tissue. It was like cutting through butter. Leaning closer, Freeman instructed Watts to take more cores. Eventually, under Freeman's guidance, Watts took a total of twelve cores from the two holes. It was not a particularly precise procedure. At one point Watts managed to get a blood vessel caught in the instrument and blood gurgled from the aperture. Still, the patient seemed OK. Nothing to worry about. An hour or so later the world's first lobotomy operation was over.

Hammatt recovered quickly. She seemed alert but much calmer. Her anxiety had disappeared; in fact she had forgotten what had caused all her problems in the first place. Hammatt could read, could name members of her family and, for the first time in months, was sleeping well. Freeman later reported how Hammatt could now manage 'home and household accounts, enjoys people, attends theatre, drives her own car'. It was wonderful. 'Great improvement,' he concludes. Then, a few months after her operation, Hammatt suffered a convulsion. It was probably related to her surgery. She fell, breaking her wrist and, according to Freeman, became 'indolent' and 'sometimes abusive'. Nevertheless, her anxiety never returned and she lived a relatively normal life.

Freeman declared the operation a great success and rushed off to tell his colleagues. When he published the details of the case, his lobotomy operation won mixed reviews. While some considered it a fine idea, others were outraged that such an untested, extreme operation was even being attempted. But Freeman was completely convinced that a surgical breakthrough had been made. Moreover, he knew how to convince others: he would ignore the medical establishment and take his radical new operation straight to the public.

In a typically long-winded headline (but with the admirable use of a semicolon) the front page of the *New York Times* proclaimed: 'Surgery Used on the Soul-Sick; Relief of Obsessions Is Reported'. The article referred to Freeman's new surgical technique as 'psychosurgery' and 'surgery of the soul'. His operation could cut away 'sick parts of the human personality' and transform 'wild animals into gentle creatures'. Out of the twenty patients Freeman had treated, the article said, 15 per cent (three) were 'greatly improved', with a further 50 per cent (ten) of them being 'moderately improved'. The article went on to detail two case histories, including Hammatt's, with only passing mention of two deaths following the procedure and unattributed criticism from some 'leading neurologists'.

The *New York Times* was not alone in trumpeting this new and exciting operation. It was proclaimed a miracle, an incredible cure and even, according to one gushing news report, 'one of the greatest scientific innovations of this generation'. For the first time in history, here was a doctor who could cure madness; heal the mind with surgery.

As ever with these things, the reality did not quite live up to the hype. Many of Freeman's early patients were soon suffering relapses. His answer was to go back and repeat the operation, gouge out more bits of brain. Other patients were suffering terrible side effects. Following their operations they were acting like children: they had to be retaught how to carry out basic functions (such as using a toilet); they were lacking in energy and self-control. It was what had happened to Phineas Gage. They were not the people they had been before the operation.

Over the next five years Freeman and Watts perfected their technique as they conducted more and more lobotomies. Other surgeons took up the procedure, while Freeman worked his way through an ever-increasing list of patients. Soon he and Watts were conducting operations on conscious patients using local anaesthetic. Freeman would have them count or sing a song so that he could tell what effect the leucotome was having. In one instance he is even said to have asked his patient to recite the Lord's Prayer – an unfortunate choice, given the circumstances.

In 1941 Freeman was approached by Joe Kennedy and asked to operate on his daughter Rosemary – the sister of future President John F. Kennedy. Strictly speaking, Rosemary was a poor candidate for a lobotomy. Quiet and beautiful, there is little evidence that there was anything much wrong with her. She might have been suffering from a learning disability, or perhaps depression. People whispered that she was not quite right in the head, and that would not do for an overachieving Kennedy. Joe Kennedy was very persuasive, so Freeman and Watts agreed to go ahead with the surgery.

The procedure was carried out in secret. Joe Kennedy did not even tell his wife. When Rosemary came round from the anaesthetic, she was a very different person. Slow and emotionless, she was hardly able to move or speak. Although she eventually learnt to walk again, she was left permanently disabled and ended up in a residential institution in Wisconsin. If anyone asked, they were told that Rosemary was suffering from a mental illness. Better than saying she had been lobotomized. Freeman never said a word about the case. It was in his best interests not to publish the details of any high-profile failures.

Despite the odd setback, everything was going well for Freeman, but he was not satisfied. Lobotomy operations were taking too long and the asylums were filling up fast. There was no way he was going to get through all the patients that needed this miracle surgery of the soul. To add to Freeman's frustration, the procedure had to be carried out by a qualified neurosurgeon, but he wanted to do it himself. He needed a way to make it simpler and faster. Up until this point he had always made sure to describe the lobotomy as surgery of 'last resort', but this was about to change.

PRODUCTION-LINE LOBOTOMIES

Washington DC, January 1946

Twenty-nine-year-old Ellen Ionesco arrived at Freeman's office accompanied by her husband and daughter. Freeman was their last hope; otherwise they feared that Ellen would have to be admitted to hospital before she killed herself. Over the past few weeks her condition had worsened. She suffered from terrible depression and would lie in bed for days. She was paranoid, suicidal and lapsed into terrifying bouts of violence. At one point Ellen had even attempted to smother her six-year-old child.

Sunlight streamed through the windows as Freeman examined the patient and carefully explained what he was planning to do. It was clear to him that she needed immediate treatment. Of course, any new procedure had its risks, but Freeman was so kind and reassuring that it did not take long for them to agree. Didn't the doctor always know best? Freeman ushered the patient through to a back room, where the equipment was already laid out and asked Ellen to lie down on the examination couch. The operation would not take long, he told her. Before she knew it she would be on her way home. He asked his secretary to order her a taxi.

Freeman slipped a rubber tube between Ellen's teeth and powered up the ECT machine. He fastened a belt containing electrodes around her head. The ECT machine hummed. He asked Mr Ionesco to help hold down his wife. Freeman flicked the switch. The electrodes crackled as Ellen convulsed on the couch, her jaw locked, her head twisting from side to side. Freeman pushed the switch again until his patient was finally rendered unconscious by the electricity. He was ready to begin.

After draping a cloth beneath her eyes, Freeman pulls back one of Ellen's upper eyelids and picks up an ice pick. It is an ordinary ice pick – the type found in many American homes. It looks like a chisel with a wooden handle, a long shaft and a strong, sharp point. It even has the company's name on the side: the Uline Ice Company.

Holding the upper eyelid in his left hand, with his right he inserts the tip of the ice pick into the top of the eye socket. He is careful not to damage her eyeball as he pushes the ice pick diagonally upwards into her tear duct, following the line of her nose.

Steadying the pick, he reaches for a hammer.

Bang, bang.

There is a crunch as the ice pick punches through the thin transorbital bone at the top of her eye socket. Freeman pushes the pick through the bone and into the frontal lobe of the brain. He wiggles the tool from side to side, slicing through the nerve tissue. He pushes it in further, sweeping it across like the blade of a windscreen wiper. After a couple of minutes, he gives the pick a final twist and yanks it out of Ellen's eye socket.

In less than ten minutes, the operation is over and the patient starts to come round. She is helped from the table but can hardly walk. She is disorientated; the eye Freeman operated on is black and blue. She looks as if she has been beaten up.

A week later, Freeman performs a second lobotomy through her other eye. In future he plans to do both eyes at once. After a few days in bed, Ellen is transformed. She is calm, her crazed mind now at peace. She takes up gardening, works in a shop and eventually trains as a nurse. Ellen Ionesco has her life back.

Freeman called his new procedure the transorbital lobotomy. It was quick and easy. It didn't need an anaesthetist, surgeon or operating theatre. There was no faffing around with antiseptics, masks or gloves. As long as he made sure the ice pick was sterilized, that was good enough. The best thing was that the whole operation was so simple that almost anyone could do it. Freeman was tremendously excited. He would be able to transform the lives of thousands of people with mental illness... he could train other doctors... it would be a new era of psychosurgery for the masses! Just as Henry Ford had invented the production-line car, so Freeman had devised the production-line lobotomy.

Dr Freeman jumped into his camper van and set off across the United States to spread the word. During the next few years he criss-crossed the country performing his transorbital lobotomies. He travelled through Europe; he visited clinics and hospitals, operating on one patient after another. As he became more adept, he began to refine the procedure. He would hold the ice pick in his left hand, even though he was right-handed, drive two ice picks in at once, and even carried out operations with a carpenter's mallet. It was wonderfully easy. On one particularly memorable day Freeman got through twenty-five patients. His operations became performances, as doctors, reporters and the occasional interested bystander watched with horrified fascination.

The sight of an ice pick being pushed into a patient's eye socket was bad enough, but the sounds were, if anything, even more gruesome: the buzz of the ECT followed by the thump of the hammer and the crack of bone, the swishing back and forth of the ice picks and the faint plop as Freeman yanked them out. As each disorientated patient staggered from the table, their black eyes smarting, the doctor could notch up another success.

Freeman was getting through an awful lot of patients. He personally performed around three and a half thousand lobotomies, and trained doctors across the world. In total, it is thought that around one hundred thousand people were lobotomized. The results were mixed. Some, like Ellen Ionesco, returned to their families to lead relatively normal lives; Ellen's daughter speaks of Freeman with affection. Other patients were not so fortunate. Following lobotomy their personalities were irrevocably changed; they became docile, placid, mindless; dead to the world around them. A slip of the ice pick left some patients paralysed after the operation. Some died from complications. But Freeman seemed blind to the failures and oblivious to criticism. By the 1960s his notion that lobotomy was a surgery of 'last resort' – to be performed only on the desperately ill – had gone completely out of the window.

THE CASE OF HOWARD DULLY

California, 1960

It was a child's worst nightmare. Howard Dully's mother died when he was five, and she was replaced by a stepmother, Lou, who never loved him; she didn't seem even to like him. She was intolerant and criticized Howard, punishing him for things he didn't do, treating him differently from his brother.

Howard was no angel. He resented his new mother's presence. He could be moody, disrespectful and argumentative, but then so can most young boys. There was nothing particularly unusual about Howard. He was a perfectly healthy boy.

Howard's father was away for long periods of time, and even when he *was* around, he appeared oblivious to what was going on. As the years progressed, the relationship between Lou and Howard deteriorated further. The boy was getting into trouble at school; his stepmother would yell or hit him for the most petty of reasons, and the boy would yell back. They were constantly arguing, slowly driving each other up the wall. Howard loathed his impostor of a mother, and Lou was fed up with her stepson. Sooner or later something had to give.

Over the years, Lou had consulted doctors, psychologists and psychiatrists (six psychiatrists in 1960 alone) about Howard. All of them assured her there was nothing wrong with him. His behaviour might sometimes be challenging, but he was perfectly normal. Even Howard's father couldn't see anything wrong. Angry, and increasingly frustrated, Lou was eventually referred to Walter Freeman, who had set up his offices in Los Altos, California. Perhaps he would find something wrong with the boy?

Howard's stepmother had her first meeting with Freeman in October 1960. She went alone. Lou told Freeman about Howard's behaviour. Some of it was true, some of it made up. She explained how when she had first seen Howard she thought he was a 'spastic' because of 'a peculiar gait' (the boy was then five and later turned out to be good at sports). She recounted how Howard didn't play with toys, but was destructive with them (much of the time Howard played on his own, so how would she know?). She told Freeman that Howard hated to wash (he was a boy!), she said he daydreamed and scowled if the TV was tuned to some programme other than one he liked, as if this were some sort of damning indictment. Tellingly, Freeman referred to the notes he made of this first meeting as 'the articles of indictment'.

The doctor's notes ran to several pages. He seemed to accept everything he was being told, including Lou's claim that Howard urinated on his bedroom wall or defecated in his trousers. She would make up anything to convince the doctor that Howard was mentally ill, and it seemed to be working. Freeman noted that the indictment was 'sufficiently impressive' and that Howard was suffering from childhood schizophrenia.

Over the coming days, in the manner of some sort of self-appointed judge, Freeman took 'evidence' from Howard's aunt, the school janitor and finally the boy's father. Much of it contradicted what Lou had said. Howard met Freeman for the first time on 26 October and, like many patients before him, felt relaxed in Freeman's presence. The doctor seemed kind and gentle, willing to listen to what the boy had to say.

Taken together, Freeman's notes paint a picture of an eleven-year-old boy living in a dysfunctional family. Today they would probably be offered counselling, but Freeman had other ideas. He told Howard's father that the boy was schizophrenic and that something needed to be done 'pretty promptly'. He offered to change Howard's personality (for a reasonable fee). The decision to operate was made on 30 November 1960 – the date of Howard's twelfth birthday.

Howard Dully was admitted to a small private hospital in San Jose on Thursday 15 December. The next day he was taken into the operating room, where he was given four jolts of electricity from the ECT machine. Freeman noted that he thought it was 'one more than necessary'. Then the doctor stuck in his 'orbitoclasts' (he had moved on from ice picks to these specially designed instruments) and jiggled them around in Howard's brain. He took a picture of the two orbitoclasts protruding from the boy's head before pulling them out again. A small amount of bloodstained fluid oozed from each bruised eye socket.

When Howard's brother saw him shortly after the operation he thought Howard looked like a zombie. The boy was listless and staring. It was as if a fog had settled across his mind. But gradually Howard started to recover. However, the operation that his stepmother had hoped would make him docile and obedient seemed to have the opposite effect. He became increasingly disruptive until his parents could take it no more. Howard was sent away, first to other people's homes, then, even though he had committed no crime, to a juvenile detention centre. Finally, he ended up in a psychiatric institution, the only child in a hospital full of mentally ill adults.

Howard has spent most of his life coming to terms with what happened to him. He suffered problems with work, relationships and money. He drifted in and out of jobs and in and out of jail. Gradually, he was able to piece his life back together. Today he holds down a job as a bus driver. There is absolutely nothing about him to suggest that he has two black holes in his brain. What saved him from going completely off the rails was probably his youth. Howard's young brain was able to rebuild neural pathways and compensate for the damage Freeman had inflicted.

Dr Freeman operated on a total of nineteen children, including a four-year-old. By the time of Howard Dully's operation in 1960, even as surgery of 'last resort' lobotomy should have been confined to the history books. Drugs were available that did much the same thing only without the danger or permanence of surgery. Some drugs were even marketed as 'chemical lobotomies'. These new treatments should have put Freeman out of a job and saved Howard Dully. They almost did. By the mid-1950s the weight of criticism was piling up and Freeman had fallen out of favour in Washington. This prompted his move to California, where he offered his transorbital lobotomies to all-comers, as a quick fix for neurotic housewives or disruptive children. Somewhere between devising the procedure in the 1930s and practising it on children, Freeman had lost sight of the reasons for developing the lobotomy operation in the first place.

Some have described Freeman as a monster, sometimes in those same newspapers that sang the praises of his 'miracle' surgery in the 1930s. Even his own son, who witnessed the terrifying spectacle of one of Freeman's transorbital lobotomies, described the operation as 'diabolical'. But it is difficult to reconcile the image of a monster with the kind and gentle doctor his patients encountered. When the lobotomy was conceived it seemed to provide the only treatment for chronic mental illness. It certainly transformed some people's lives for the better.

But Freeman's greatest failure of judgement was not knowing when to stop. Nor, to be fair, did anyone step in to stop him. Where, for instance, were the authorities who should have prevented the operation on Howard Dully? The fact that Freeman kept performing transorbital lobotomies when the procedure was discredited and opposed by almost the entire medical establishment makes it difficult to forgive him for what he did to so many people. Rosemary Kennedy, Howard Dully and hundreds of others would have had very different lives were it not for Walter Freeman.

Freeman performed his last transorbital lobotomy in 1967. He was seventy-two. His patient suffered a haemorrhage and died three days later. The hospital where he was operating finally decided that enough was enough and stopped him from performing any further lobotomies. It was the end of his career. The lobotomist had lost his purpose in life. But rather than stay at home, he got back in his camper van and headed off for one final road trip.

Over the next few years he covered some fifty thousand miles, tracking down his former patients. It was as if he were seeking redemption. Perhaps he was beginning to have doubts about his treatments and wanted to prove that he had helped people, had improved his patients' lives. He visited homes and hospitals; he saw people who were very sick and those whose lives had undoubtedly been changed for the better.

Walter Freeman died in 1972. The transorbital lobotomy died with him, but to the end of his life he believed in what he had done, and he believed it was right.

Freeman was not completely misguided. Lobotomies are no longer performed, but psychosurgery – using surgery as a treatment for mental illness – is still practised in hospitals around the world. Once again, it has become a treatment of last resort. Although the lobotomy is undoubtedly one of the most disturbing operations in the entire history of surgery, when it comes to matters of the mind it is by no means the only controversial therapy.

CONTROLLING THE MIND

Córdoba, Spain, 1964

Dr José Delgado was brave. There is no doubt about that. Few scientists would attempt to do what he was about to do. Delgado claimed he knew a little bit about bullfighting, and was also reasonably confident that he knew about the workings of the brain. But combining the two? That was a different thing altogether.

The bull is charging across the ring towards Delgado, its nostrils steaming as it kicks up the dirt in the arena. Bred to be aggressive, it is an awfully big, angry bull, its head bowed down as it charges, its horns hard and sharp. If Delgado gets this wrong, there is a good chance he'll be killed.

Delgado concentrates as he grasps the matador's red cape in his right hand. He stands by a wooden barrier that he can duck behind if the experiment goes wrong, but there is no guarantee he will make it in time. In his left hand he holds what looks like a transistor radio – a small box with a long aerial extending from the top. If anything, Delgado is holding this tighter than the cape.

The bull's hoofs pound. Delgado stands his ground. The bull is getting closer. Delgado remains still. Ten feet, five feet. The bull is almost on top of him. Then Delgado presses a button on the box. The bull stops in its tracks, turns round and wanders away. It is an audacious stunt: a demonstration of the power of technology.

The previous day Delgado had anaesthetized the bull and implanted electrodes in its brain. When he pressed the button on the remote control he had been holding, it sent a signal to a receiver on the bull's head. This stimulated the electrodes deep inside the brain, changing the bull's normally violent behaviour. At the flick of a switch, Delgado had control of a ferocious animal.

By the 1960s, developments in electronics and a better scientific understanding of the workings of the brain were making such impressive demonstrations possible. Delgado used an implant device consisting of a sealed capsule about the size of a small watch. This contained the receiver and all the electronics. Wires emerged from the side of the capsule and these were sunk into particular regions of the brain. Thanks to advances in the understanding of brain anatomy and localization, Delgado could surgically implant the electrodes in specific places to get particular responses.

The doctor's other experiments were equally impressive, although somewhat less dangerous. Implants in the brain of a monkey could be stimulated to control the diameter of the pupil in the animal's eye. When the doctor pushed the button on the remote, the pupil would contract and dilate – it was almost like controlling the aperture of a camera. He could get other monkeys to yawn on demand or even perform a complex sequence of movements. He also experimented on cats. He could induce expressions of rage – an electrical stimulation caused them to hiss and bare their teeth. The cats even learnt to turn off the stimulation by rotating a wheel Delgado had installed in their cages.

It was the experiments on groups of animals that revealed the potential of the technology. Delgado discovered that he could use implants to control aggression in monkeys. He could also get other monkeys to do the controlling. In one demonstration he installed electrodes in the brain of Ali, the boss of a small monkey colony. When Ali's brain was stimulated it inhibited his normally aggressive behaviour. Delgado installed a lever in the experimental cage the monkeys lived in. When the lever was pressed, it activated the electrodes. Soon a passive monkey named Elsa learnt that pressing the lever could stop Ali's aggression. She now had control over Ali. Whenever he threatened her, she pushed the lever and Ali stopped. Elsa had become the boss.

It was one thing to operate on bulls, monkeys or cats, but what would happen if electrodes were implanted in human brains? Could human behaviour be controlled in the same way? Delgado hoped to use the technology to help patients suffering from severe mental illness, epilepsy or chronic pain. Rather than change behaviour by removing bits of the brain as the lobotomists had done (Delgado found the idea of lobotomy abhorrent), he planned to insert his devices in the brain instead.

When he tried the technology on humans the results were astonishing. With receivers attached to people's heads, he found he could stimulate a whole range of human emotions from fear through lust, hilarity and rage. One of his researchers is said to have narrowly avoided serious injury during an experiment conducted on a young woman with epilepsy. She was playing the guitar when the device was activated and promptly flew into a rage, throwing her guitar against the wall.

Soon other doctors were taking up Delgado's work. In 1965 psychiatrist Frank Ervin and neurosurgeon Vernon Mark tried out the technology on a sixty-three-year-old man dying of cancer. The patient was in the most terrible pain, and surgeons had run out of treatment options. Injections of morphine were no longer doing the trick, and what remained of his life was a living hell.

After drilling a series of holes in the man's head, Ervin and Mark inserted electrodes into carefully selected sites in his brain. The electrodes were connected to a plug in the patient's scalp, which led to a 'pain box' – a controller that the man carried in his pocket. When the pain got too much, all he had to do was push a button and the agony would go away. If he kept the switch pressed for more than forty-five minutes, the pain subsided for up to eight hours, allowing him to get a good night's sleep for the first time in months. The only side effect was that the device made him feel a little drunk. But then, the patient figured, if you're dying of cancer, that's not such a bad thing.

Few would argue against helping a dying patient, but the development of devices that could change people's behaviour at the flick of a switch had far more sinister implications. Delgado was well aware of the potential of his technology, but downplayed any suggestions that it might be used by sinister forces to control people's minds. In his 1969 book *Physical Control of the Mind* he dismissed the idea of an evil dictator standing at a master control centre stimulating the minds of an enslaved people. But then he didn't reckon on the CIA.

In the mid-1950s the US intelligence agency had started toying around with the idea of brainwashing individuals, invariably communists. They had looked at using hypnotism or drugs, and investigated the notion of using lobotomy to control antisocial behaviour. A psychiatrist, Henry Laughlin, had even been dispatched to witness some of Walter Freeman's operations. In his report, Laughlin suggested that lobotomy might be used as a 'neutralizing weapon' to 'quench crusading spirits' or 'zealous and fanatic communists'. If lobotomy could do that, just think what brain implants could do.

Following the Detroit riots in 1967 and subsequent civil unrest in deprived inner-city areas across the United States, Ervin and Mark suggested that brain implants might be used to subdue black rioters. In an article in the prestigious *Journal of the American Medical Association* they proposed that urban riots and other 'acts of senseless violence' could be prevented by surgery.

In 1972 psychiatrist Robert Heath came up with another idea. Why not use brain implants to 'cure' homosexuality? With a long list of ethically dubious research behind him, Heath conducted an experiment on a gay man. The psychiatrist placed electrodes inside the subject's brain and stimulated them while the man had sex with a female prostitute. The aim of the experiment was to condition the subject to want to have sex with women rather than men.

The backlash against using brain stimulation was swift. Delgado was lumped in with Ervin, Mark and Heath, even though their experiments were nothing to do with him. The rise of the implant in science fiction did nothing to help his cause. More and more conspiracy theorists claimed that the government had secretly implanted chips in their brains, and books such as *The Terminal Man* (see Further Reading) painted a terrifying portrait of a medical experiment gone wrong. That its author, Michael Crichton, was once one of Ervin's students probably did little to help.*

It looked as though it was the end of the line for the brain implant. The research was discredited, ethical approval for brain stimulation experiments became impossible to obtain, and funding for studies simply fizzled out. However, unlike Freeman's trans-orbital lobotomies, brain implants still held enormous potential.

STUART'S STORY

London, 2006

Sixty-eight-year-old Stuart Carter felt trapped within his own mind. He could think about moving but his body would not move. He

* *The reality was that if the government really wanted to control the population through mind control, they would stick drugs in the drinking water, rather than go through the impractical and messy business of implanting chips.*

could not smile or laugh or frown or cry. He had lost his sense of balance and could barely stand. When he placed his hand down on a table he could no longer lift it up. His face was an expressionless blank. He was a living statue watching life through a mask.

Carter had Parkinson's disease. This degenerative illness affects nerve cells in the brain that coordinate movement. Whereas many people with Parkinson's have uncontrollable tremors, Carter's body froze. He found that his facial muscles would no longer react to emotion and that his body would lock in a particular position. Sometimes it took him an enormous mental effort to get moving again. He knew that without treatment his condition would only get worse, but the drugs he was taking were no longer working. His mind remained sharp, but his body was gradually shutting down.

In 2006 Carter was offered the chance of having electrodes implanted in his brain. In his case, the technology, known as deep brain stimulation, would act as a jamming device. The implants would block the nerve signals in the sick areas of his brain that were causing him so many problems. The operation was not going to be easy – surgeons needed to implant the devices at the base of the brain, in an area called the basal ganglia. As it would be impossible to cut open Carter's brain, the surgeons at the National Hospital for Neurology and Neurosurgery would be operating 'blind'.

With a brace resembling a medieval torture device clamped to his head, Carter lies on the table in the bright glare of the operating theatre lamps. His head is marked with dots, and a large plastic sheet is stretched across his balding scalp. He stares at the ceiling as the final preparations are being made by the androgynous gowned and masked figures around him. In the twenty-first century lying on an operating table is still a nerve-racking experience, made worse if you know you are going to be conscious when a man you barely know starts drilling holes into your head. An injection of local anaesthetic and surgeon Marwan Hariz is ready to start.

Hariz presses a scalpel into the top of Carter's head, slicing deep to the bone. 'No pain, just noise,' explains the surgeon as he holds an oversized dentist's drill to the incision. A thin trickle of blood and tissue is sucked away with a tube as the drill penetrates deeper. A syringe-like device is clamped over the aperture and a wire implant is lowered into Carter's brain. Hariz guides the implant deep through the folds and ridges of his patient's brain, careful to avoid damage.

'You might feel something and you might feel nothing.' This is the moment of truth. 'Whatever you feel, tell us.' The surgeon presses the button to activate the implants. 'No special sensation?' Carter starts to feel a tingling in his left foot. It's working – Hariz has got the implants in the right place. Carter's brain has been electronically enhanced.

Two years later Stuart Carter can walk, laugh and smile. When he puts his hand down on a table he can pick it up again. He is free from the mental trap that Parkinson's imposed. If you look carefully at the top of his head, you can see two small bumps. You can even follow the path of the wire leading away under his skin and down his neck. It terminates in a flat box beneath his shoulder – the stimulator that sends electrical pulses to his brain.

Today there are tens of thousands of people walking around with brain implants. Some, like Carter, have electrodes deep within the brain to stimulate or block nerve signals. Many others have cochlea implants in their ears to cure deafness. Scientists are developing implants to enhance memory, to help the blind to see, or simply to connect our 'wetware' to computer software. Delgado's vision of using electronic devices in the brain to combat disease and disability is finally being realized.

Advances in brain surgery over the past 150 years have come at tremendous human cost. From patients bleeding to death on the operating table to victims of Freeman's crude lobotomies and brain implants to 'cure' homosexuality, the history of brain surgery is full of examples where surgeons have strayed well beyond the bounds of what is morally or ethically acceptable. It is still difficult to read Freeman's notes on Howard Dully without becoming angry, or contemplate Ervin's and Mark's ideas on mind control without feeling utter revulsion.

THEN AND NOW

If you look back through the history of surgery, there are some remarkable individuals: pioneering surgeons, such as Ambroise Paré, Ignaz Semmelweis and Joseph Lister, who helped make operating safe; surgeons such as Dwight Harken, Walter Lillehei and Harold Gillies – brave men who had the courage to fail; but there is also a third group – surgeons such as Walter Freeman and Alexis Carrel – who seemed to operate in their own peculiar moral universe.

Whether these people were brilliant, courageous or misguided, the history of surgery is full of individuals who were prepared to have a go to see what happens. Sometimes they succeeded, sometimes they failed. Some patients lived, many others died. But even in death, they will have helped future patients to live.

Without the surgeons who stole corpses from gibbets, who cut into beating human hearts, or who first attempted to operate on the human brain, modern surgery – from routine procedures to repair a damaged knee to hi-tech operations to implant a device in the brain – would not have been possible.

Nevertheless, despite all the advances, there are still many things that challenge twenty-first-century surgery: artificial hearts are cumbersome, organs still get rejected, reconstructive surgery cannot always restore a face, a brain tumour can still kill. Even the most minor surgery is not without risk: hospital patients still die of major infections; anaesthetics are not 100 per cent reliable; the knife can still slip.

Most of us know people who have gone into hospital for an operation and never come out. Surgeons are remarkable people but, despite the impression some of them might give, they are not gods. There are good surgeons and bad surgeons, but even the best surgeons are fallible. When we submit to an operation we put an immense amount of trust in those who wield the knife. In most cases our trust is fully justified, but the history of surgery suggests that sometimes it is not.

If you need an operation, just be grateful that you are alive today and not 170 years ago – the next patient on Robert Liston's operating schedule.

TIMELINE

10,000 BC	Evidence that trepanning was in widespread use.
1500 BC	First recorded plastic surgery (nasal reconstruction).
c. AD 157	Galen appointed as surgeon to the gladiators at Pergamum in western Turkey.
1536	Vesalius acquires a skeleton and begins to unravel human anatomy.
1545	Ambroise Paré publishes his first *Treatise on Gunshot Wounds.*
1597	Gaspare Tagliacozzi publishes the first-ever book on reconstructive surgery.
1765	Tooth transplants from living donors are popular.
1834	Robert Liston is appointed professor of clinical surgery at University College Hospital, London.
1846	The first operation is carried out with ether anaesthetic.
1847	Ignaz Semmelweis successfully combats puerperal fever.
1847	James Simpson develops the chloroform anaesthetic.
1848	The first death from chloroform is recorded.
1848	Phineas Gage survives having an iron rod fired through his head.
1857	Louis Pasteur discovers that germs cause living matter to decay.

1865 Joseph Lister treats a patient using 'antiseptic' techniques.

1884 Rickman Godlee and A. Hughes Bennett perform 'successful' brain surgery.

1894 Assassination of French president Sadi Carnot and start of Alexis Carrel's career.

1902 Luther Hill conducts successful heart surgery.

1903 Gladys Deacon has paraffin wax injection.

1912 Alexis Carrel wins Nobel prize.

1917 Harold Gillies performs facial reconstruction operations and develops tube pedicle.

1931 Harvey Cushing conducts his two thousandth brain tumour operation.

1935 Carrel publishes his book on the future of humanity, *Man, the Unknown.*

1935 Egas Moniz performs first 'leucotomy' operation.

1936 Walter Freeman and James Watts carry out first lobotomy operation.

1939 Archibald McIndoe begins treating Allied casualties.

1943 Willem Kolff invents a successful dialysis machine.

1944 Dwight Harken cuts into a beating human heart.

1944 Death of Alexis Carrel.

1946 First female to male sex change operation.

1946 Walter Freeman performs the first transorbital lobotomy.

1951 Paris surgeons attempt kidney transplant with organs from executed criminals.

1952 F. John Lewis carries out the first successful open-heart surgery.

1953 John Gibbon performs successful operation using heart-lung machine.

1954	Walter Lillehei conducts first cross-circulation operation.
1954	First successful kidney transplant (between two identical twins).
1955	Lillehei and Dick DeWall invent a reliable heart-lung machine.
1955	Denis Melrose develops a method of stopping and starting the heart during surgery.
1957	Surgeons in Boston use radiation to destroy the immune system in transplant patients with limited success.
1958	Denis Melrose performs open-heart surgery on live television.
1960	Howard Dully is lobotomized by Walter Freeman.
1964	José Delgado stops a rampaging bull by remote control.
1967	World's first heart transplant performed by Christiaan Barnard.
1967	Freeman performs his last lobotomy.
1970	Frank Ervin and Vernon Mark propose using brain implants to suppress violent tendencies in black rioters.
1976	Roy Calne begins experiments on cyclosporine.
1978	First successful transplants using cyclosporine.
1998	Clint Hallam receives the world's first hand transplant.
2001	Clint Hallam's hand is amputated.
2005	First successful (partial) face transplant.

FURTHER READING

I was surprised when I started writing *Blood and Guts* how few books are available on the history of surgery. There are some excellent books on medicine, particular surgeons or episodes, but few broader histories. If you want to find out more about a particular era, surgeon or practice (such as lobotomy), I have listed some suggestions below. A few of the books are out of print, but, thanks to the wonders of the Web, are usually obtainable. Rather than give a long list of my sources and references for this book, which would run to a separate, sprawling and dull chapter, I thought it might be helpful to point to some general sources of surgical history, facts and inspiration.

I cannot recommend highly enough a visit to the Old Operating Theatre Museum and Herb Garret near London Bridge (www.thegarret.org.uk). Here, up a narrow winding staircase at the top of a church, is the original operating theatre from St Thomas' Hospital, complete with operating table and surgical instruments from the early nineteenth century. The museum also has displays on early medicines, bleeding and the development of anaesthetics and antiseptics. One of my favourite exhibits is a walking stick that a surgeon had his patients bite on during operations, to help them cope with the pain. You can still make out the tooth impressions.

Another fascinating museum is the Hunterian at the Royal College of Surgeons near Lincolns Inn (www.rcseng.ac.uk/museums). I have not written much about John Hunter (see Chapter 3), but in the museum you will find an incredible collection of the weird and wonderful. You will also see how much he achieved in advancing our understanding of human biology.

If there is one single institution that made this book possible, it is the Wellcome Library (on Euston Road in London). I have spent many happy hours there, leafing through old books, papers and journals. The library contains Robert Liston's books on Victorian surgery and the case notes for the first anaesthetic. It holds translations from Vesalius and Semmelweis, papers on the first heart surgery, and even descriptions of groundhogs (see Chapter 2). The biggest problem with the Wellcome Library is that it is very easy to get diverted. I spent an afternoon reading graphic case notes on early eye operations, only to realize that there was no space for them in the book. The library is free to join and much of it (particularly the images) is accessible online (http://library.wellcome.ac.uk).

The library is housed alongside the Wellcome Collection, where you will find beautifully displayed surgical knives, cupping bowls and an exhibition on the latest developments in biotechnology.

The book that I think gives the most complete account of surgical history is *A History of Surgery* by Harold Ellis (Greenwich Medical Media, 2000). Ellis is a world expert on surgical history, and although the book was probably aimed at a medical and surgical readership, it is clearly written and well illustrated. Unfortunately, the book is out of print, but I understand a new edition is in the pipeline. For a glimpse into the mind of a surgeon I recommend Atul Gawande's thrilling and entertaining *Complications: A Surgeon's Notes on an Imperfect Science* (Profile, 2002). In a similar vein David Wootton's *Bad Medicine: Doctors Doing Harm since Hippocrates* (Oxford University Press, 2006) suggests that, until relatively recently, doctors were often doing more harm than good.

I also relied on several human physiology and anatomy textbooks. I hope that, with their help, I have not made any glaring errors in my anatomical explanations. The other book that proved invaluable was my mum's nursing textbook from the 1940s, *A Complete System of Nursing* (Temple Press, 1947) which gave useful information on the treatment of patients and an alarming insight into just how basic medical practice was even then (penicillin is mentioned only briefly as it was not yet widely available, and anaesthetics were still administered from a 'drop bottle' on to a mask).

Below are a few further reading suggestions:

The Greatest Benefit to Mankind, Roy Porter (HarperCollins, 1997)
This substantial book covers the whole history of medicine, but is readable throughout. Porter has written many books on medical history, all of which are equally impressive.

Moments of Truth: Four Creators of Modern Medicine, Thomas Dormandy (John Wiley, 2003)
Stories about four of the people who helped shape modern medicine, including an excellent section on Ignaz Semmelweis.

Seven Wonders of the Industrial World, Deborah Cadbury (Fourth Estate, 2003)
Based on the BBC series of the same title, it includes a section on John Snow, and in the chapter on the transcontinental railroad gives an excellent account of the environment that Phineas Gage (see Chapter 5) would have worked in.

King of Hearts, G. Wayne Miller (Crown, 2000)
Although the writing style is sometimes a bit sentimental, this is nonetheless a gripping biography of Walter Lillehei.

The Knife Man, Wendy Moore (Bantam Press, 2005)
An extremely entertaining and evocative biography of John Hunter.

Transplant: From Myth to Reality, Nicholas L. Tilney (Yale University Press, 2003)
This is a rather skewed account of transplant surgery from a US perspective, but it does give an insider's overview of the discipline's development. Far better, I think, is Joseph E. Murray's *Story of the First Human Kidney Transplant* (Mitchell Lane, 2002).

There are three superb books relevant to the chapter on plastic surgery: *Gillies: Surgeon Extraordinary*, Reginald Pound (Michael Joseph, 1964); *Gladys, Duchess of Marlborough,* Hugo Vickers (Weidenfeld & Nicolson, 1979) – a fascinating account of an extraordinary life; *The Reconstruction of Warriors*, E.R. Mayhew (Greenhill Books, 2004). I also recommend taking a look at Jacqueline Saburido's website (www.helpjacqui.com).

An Odd Kind of Fame: Stories of Phineas Gage, Malcolm Macmillan (MIT Press, 2000)
This is a forensic account of Gage and the myths surrounding him. The book is meticulously researched, but does come from an academic perspective, so it can be a bit dry at times. However, Macmillan's research in uncovering the story of Gage is truly impressive.

Harvey Cushing: A Life in Surgery, Michael Bliss (OUP, 2005)
There is so much more to say about Cushing than I have been able to, and it can be found in this well-written and accessible account of his brilliant and complex life.

The Lobotomist, Jack El-Hai (John Wiley, 2005)
If you read only one other book (apart from mine) on the history of surgery, make it this one. It is superbly researched and entertainingly written. If you want more, get a copy of *My Lobotomy* by Howard Dully (Vermilion, 2008). It includes some of Freeman's original notes and Dully's personal journey to try to understand why he was lobotomized.

The Terminal Man, Michael Crichton (Arrow Books, 1972)
Although this is a work of fiction, in the light of some of the experiments being conducted at the time, it turns out to be terrifyingly close to reality.

Finally, if you want an insight into the, shall we say, hinterland of medical research, I suggest taking a look at Alexis Carrel's *Man the Unknown* (Harper & Bros, 1935) and *The Culture of Organs* (P.B. Hoeber, 1938). Likewise, José Delgado's *Physical Control of the Mind* (Harper & Row, 1969) is worth flicking through, if only for the pictures.

INDEX